Conference sponsored jointly by the Institution of Civil Engineers and the Institution of Structural Engineers

ORGANIZING COMMITTEE

W. G. Curtin (Chairman)
R. E. Bradshaw
R. G. D. Brown
J. Parkinson
A. K. Tovey

CONVERSION FACTORS

1 in	0.0254 m
1 ft	0.3048 m
1 UK ton	1.016 t
1 lbf	4.448 N
1 lbf/in^2	6895 Pa

Published for the Institution of Civil Engineers by Thomas Telford Ltd, PO Box 101, 26–34 Old Street, London EC1P 1JH

British Library Cataloguing in Publication data:

Reinforced and prestressed masonry.
1. Prestressed construction—Congresses
2. Masonry—Congresses
I. Institution of Civil Engineers
II. Institution of Structural Engineers
624.1'83 TA425

ISBN: 0 7277 0161 4

Printed by The Thetford Press Ltd, Thetford, Norfolk.

Contents

1. Development, application and potential of reinforced and prestressed masonry. W. G. CURTIN 1
2. Reinforced brickwork in the George Armitage Office Block, Robin Hood, Wakefield. R. E. BRADSHAW and J. P. DRINKWATER 13
3. Examples of the application of reinforced blockwork. S. ADAMS and J. E. SAUL 23
4. Reinforced masonry cantilever construction. R. J. M. SUTHERLAND 31
5. Grain silos in reinforced brickwork. R. BEARD 43
6. Design and construction of a reinforced brickwork tank. G. D. JOHNSON 53
7. Reinforced brickwork in the Hyde Park Methodist Church, Leeds. R. E. BRADSHAW 69
8. Post-tensioned, free cantilever diaphragm wall project. W. G. CURTIN, G. SHAW, J. K. BECK and L. S. POPE 79
9. Reinforced and prestressed masonry in agriculture. J. P. DRINKWATER and R. E. BRADSHAW 89
10. Prestressed blockwork silos. T. J. S. MALLAGH 97
11. Post-tensioned brickwork diaphragm subject to severe mining settlement. G. SHAW 103
12. Codification of reinforced and prestressed masonry design and construction. B. A. HASELTINE 115

Discussion 123

Closing address. C. J. EVANS 153

Bibliography 155

1. Development, application and potential of reinforced and prestressed masonry

W. G. CURTIN, MEng, PhD, FICE, FIStructE, MConsE, Consultant, W. G. Curtin and Partners, Liverpool, and Royal Society Industrial Research Fellow, UMIST

1.0 INTRODUCTION

It is patently obvious that since brick and blockwork are strong in compression and weak in tension that, like concrete, they can be reinforced to carry tensile stresses or prestressed to eliminate them. What has not been obvious is that the application of modern structural techniques and developments (in other materials) to normal brick and blockwork makes them an economic and fast method of construction. We have a 'new' structural material. Reinforcing and prestressing widens the application, improves the competitiveness and increases their potential.

A common attitude amongst some engineers is that brickwork is merely a non-structural material for houses or cladding and that blocks are only a cheap substitute for bricks. Perhaps not surprisingly this view is, on the whole, shared by the brick and block industry who tend to regard their material as a waterproof 'wall-paper' to wrap around structures (or to sub-divide them) since in the life-time of the people in the industry that has been the major outlet for their product. But this attitude wastes the material, makes for uneconomic construction and severely limits the potential application.

Many structural designers first lay out a steel or concrete frame and then add walls to keep out the weather, enclose lifts and staircases, sub-divide the space etc. In many cases if the walls were designed first there would be no need for a frame. Frequently retaining walls are automatically designed in reinforced concrete without considering the alternatives. Many other examples could be cited where engineers have not considered, or appreciated, the application of the 'new materials - the purpose of this symposium is to show such applications.

2.0 DEVELOPMENT

This is no place to unfold a long saga of the history of the developments but it may be appropriate to show briefly that they are not new,risky or suspect, but are well-tried and proven in practice. In fact reinforced brickwork pre-dated reinforced concrete. One of the known early applications was by Brunel on the shafts at Rotherhithe for the Blackwall Tunnel in the 1820's. But the introduction of firstly cast-iron then structural steel and finally the reinforced concrete frame practically wiped brickwork out of the structural market in the industrialised countries. However, countries like India and Japan, continued to use structural brickwork because for them it was much cheaper than steel or concrete. Since both countries had earthquake problems there was a greater need for reinforced brickwork. The Quetta earthquake gave it a particular boost. There was a revival of interest, for the same reason, in California along the San Andreas fault-line. In England in the immediate post-war period there was a structural steel 'famine' and many contractors and some engineers lacked the confidence or experience to use reinforced concrete. They chose reinforced brickwork - sometimes with unhappy results due to lack of cover to the reinforcement, separation of grout lifts by mortar droppings etc. Then the easing of the steel famine together with the education drive of the CACA resulted in a decline of interest in reinforced brickwork- and blockwork had hardly started to make any impact.

The serious revival of structural masonry did not occur in Europe until the '50's when the Swiss engineer Haller built some spectacular high-rise flats - which passed almost unnoticed by British engineers. This work was followed in the late '50's and early '60's by a handful of British engineers who, being busily occupied in producing structures, did not produce papers - the hallmark of 'success' and 'respectability' in some research and academic fields. (It has not been easy to collect for this symposium papers from knowledgable engineers heavily engaged in designing and building - to paraphrase Shaw's unfair criticism of teachers "engineers who can, do, - those who can't, write papers". This symposium proves that this rule, too, has exceptions!) It was soon appreciated by these pioneers that the application of the concrete techniques of reinforcing and prestressing could be extended to masonry.

The structural use of concrete blocks has developed more recently than structural brickwork. This may be due to the fact that they are of relevantly recent innovation, the early blocks were of low strength,had a reputation for unreliability and were thought of as a low-cost substitute for brick cladding. This situation could well change rapidly. Concrete blocks are now reliable, have higher strengths and, possibly even more important, the concrete research and development organisations have far more experienced and practised engineers than the brick counterpart. It is interesting to note that the brick industry has not a single experienced practical engineer on its senior staff - a situation that may not be conducive to rapid progress). Since concrete is, psychologically, a more readily acceptable material to the majority of engineers than brickwork it is possible that the advance in concrete blockwork may be more rapid than in clay brickwork. (Already a greater area of walling is now constructed per annum in concrete masonry than in brick masonry. This is not unexpected when it is realised that the bulk of masonry construction is non-structural and thus does not utilise brickwork's structural properties).

3.0 APPLICATIONS

3.1 INTRODUCTION

The applications fall into two main groups:-

(i) To improve the vertical load-bearing capacity and

(ii) to improve the bending resistance.

There are many other applications such as:-

(iii) provision of resistance to accidental damage

(iv) provision of resistance to in-plane tensile stresses in walls subject to differential settlement, racking shear etc.

(v) cover (and infill) to reinforcement in hoop tension for silos etc.

(vi) increasing resistance to tensile stresses in arches subject to asymetrical loading (rolling loads etc).

(vii) just sheer downright fun on the part of the engineer and his client - and why not?

3.2 Improvement of Vertical load-bearing capacity

Main reinforcement is added to concrete columns to increase their load resistance and the same basic principle has been applied to masonry. A column of hollow concrete blocks is basically a permanent shutter which when reinforced and grouted up adds to the strength of the final composition. This obvious and simple application can also be used in brickwork when there are restrictions on the overall size of the column.

Hollow concrete blocks can thus be considered as permanent shutter to in-situ reinforced concrete and bricks as large aggregates. This almost childish exposition has been found helpful in training graduate engineers who have been taught nothing about plain masonry let alone reinforced or prestressed!

Walls too can be strengthened in compression by reinforcing though this is rare since it is easier to use higher strength bricks or blocks or change the geometric configuration of the wall by adopting such shapes as the diaphragm, fin etc. When planning restrictions prevent this then the simple alternative is to add reinforcement.

3.3 Improvement of lateral load-bearing capacity - increased bending resistance.

3.3.1 Walls. Structural masonry is mainly a 'wall' material and is not often an economic 'beam' material. Probably the most common application of the technique is for retaining walls.

3.3.2 Concrete block retaining walls - reinforced The cellular block wall is the classic case of the precast permanent shutter and there is little doubt that the use of this technique will become more wide-spread as engineers become more familiar and experienced in its application.

3.3.3 Reinforced brick retaining walls. There are a number

of established methods of reinforcing brick walls,the more common are outlined below:-

(i) Grouted cavity wall is probably the simplest. The main and secondary reinforcement is fixed, the two leaves of brickwork constructed and the cavity grouted.

(ii) Quetta bond has been popular and amongst some engineers still is. The author has found that contractors are not over-enthusiastic and have described it as a 3-D jigsaw Puzzle. Grouting up using the bricklayers mortar must be carried out simultaneously with bricklaying.

(iii) Pocket or composite construction for low-height walls. A solid wall is constructed leaving pockets which are later reinforced and concreted. The result is a line of reinforced concrete columns with brickwork panels spanning horizontally between them.

3.3.4 Blast - resistant walls.

After the Flixborough disaster it is now mandatory to provide blast-resistant walls in structures which might be subject to the effect of explosion. As far as can be ascertained this requirement appears to have been met in structural masonry by the use of reinforced cellular block walls.

3.3.5 Wind Resistant walls.

The walls of tall single-storey structures (factories, warehouses etc.) are rarely subject to significant direct compressive stress and their design is mainly governed by the necessary resistance to the lateral pressure due to wind. Even when the masonry is used as mere cladding and not structurally, as it should be, it is still necessary to provide such resistance(- or watch the wall collapse in the first strong gale!) When the wall is used more structurally efficiently, as for example in fin and diaphragm construction, the lateral pressure resistance can be further enhanced by prestressing or reinforcing.

3.3.6 Beams

Though beams can be made with masonry - and are so made for aesthetic prestige and other reasons as yet they are unlikely to prove cost competitive against the present alternatives. The main application has been in lintols over door openings etc. and it is in such applications that reinforced blockwork has proved - for the author, to be much simpler than brickwork. (Though recent research work at UMIST on post-tensioned brick box-beams looks promising).

3.3.7 Wall Beams.

Reinforcing the lower courses of walls has long been a method of creating a 'beam' which is resistant to differential settlement. This method was recommended years ago for housing in mining areas.

4.0 REINFORCE OR PRESTRESS ?

The decision to reinforce or prestress bricks or blocks will depend mainly on economic considerations, whether bricks or blocks are predominant in the structure and on aesthetic grounds.

As a crude rough guide it would appear, to the author, preferable to prestress brickwork and to reinforce blockwork - though of course, as always, there will be some occasions when the reverse holds good.

The preference for reinforced blockwork is because:-

i) blocks form a permanent shutter.
ii) generally it is faster to construct and with lower labour input than brickwork.
iii) it is simpler (and more certain) to grout up.

The preference for prestressed brickwork is because:-

i) bricks generally have a higher compressive strength than blocks.
ii) it is simpler to form sections with a high Z/A ratio and radius of gyration in brickwork.
iii) brickwork is more likely to be adversely affected by tensile cracking at the brick/mortar interface than blockwork. Prestressing can eliminate tensile cracking and thus prestressed brickwork has increased serviceability.
iv) Reinforced brickwork beams can often waste the material's compressive strength. In the 3 brick deep beam shown in Fig.1 only the top brick is used structurally, the bottom brick provides some cover and the middle brick only ensures composite action between the reinforcement and the sole structural brick.

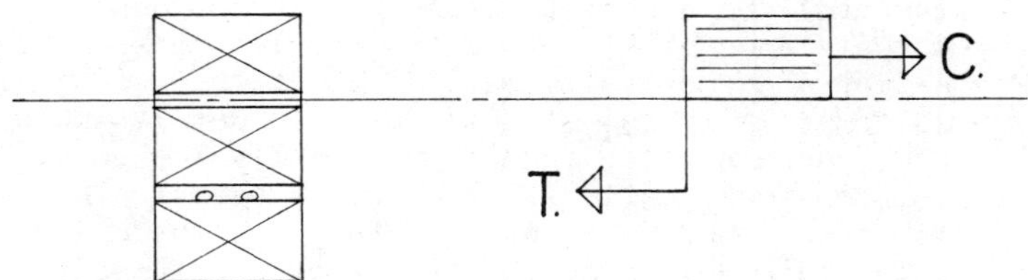

Fig. 1

Whereas in the post-tensioned box-beams shown in Fig. 2 all the brickwork is used structurally

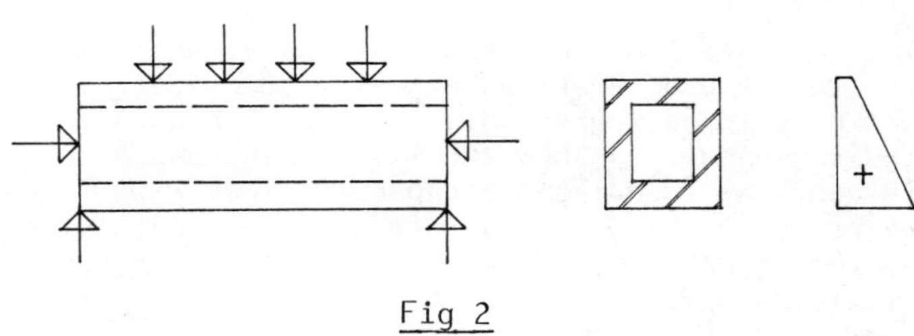

Fig 2

The chances of more than 50% of the members of this symposium agreeing with the above are probably remote! At least such a nailing of the colours to the mast should provoke discussion - and to continue the warring flavour (and cliches) the author risks being hoisted by his own petard!

4.0 'Change' in Material

To illustrate the change in the structural behaviour of this 'new' material, used in modern method, it is interesting to compare s olid brick wall and a post-tensioned diaphragm wall of the same cross-sectional area. The moment of resistance in both is f x Z (stress x section modulus). It can be shown that a diaphragm wall with a 200 mm wide cavity has a Z of about six times the magnitude of a solid wall of the same amount of material. The compressive strength of masonry is around twenty times its tensile strength; halving this gives a precompression of ten times the tensile strength (so that the combined prestress plus bending stress does not exceed the compressive strength). So a post-tensioned diaphragm has a moment of resistance of six times ten (=60!) greater than that of a normal solid wall. Too, if the solid wall was temporarily and slightly over-loaded it would crack, permanently, and be permanently weakened. The post-tensioned wall, if temporarily over-loaded, would also crack. But on removal of the load the crack could close up, under the effect of prestressing, and the wall would be as good as new. So that not only has there been a dramatic increase in strength , the material has changed from being brittle and is on the way to being ductile.

A similar exercise can be carried out to show the equally dramatic effect of reinforcing a hollow-block concrete wall.

The 'calculation' above has been heavily simplified for the sake of brevity but detailed calculations are unlikely to show gross distortions. Designers can do their own sums to show the massive increases possible in vertical load-bearing capacity. They will find that we really do have a 'new' structural material that is no longer brittle and weak in tension but strong and highly cost competitive.

5.0 GUIDANCE, CODE OF PRACTICE AND RESEARCH

5.1 Guidance

There is an excellent American textbook (ref 1) (whose even greater value to the octogenarians is that is uses Imperial measurement and the working stress philosopy!). It is not too difficult to convert from Imperial to S.1 units or from working stress to limit state. There are good guides from both the CACA (ref 2) and the BDA (ref 3) and more are promised and there is an English book available (ref 4). Guidance though no matter how good is no substitute for engineering judgment and experience - this is in no way to suggest any denigration whatsoever of the above publications for a wise man can learn from the experience of others - only fools learn solely by their own mistakes.

5.2 Code of Practice

The B.S.I's Draft Code Part 2 'Reinforced and Prestressed Masonry' (ref 5) was issued for public comment last year - and received very little from practising engineers! Whether this was due to agreement with the draft, lack of interest in it or other reasons is not certain. Sixteen pages of the draft are devoted to the 'design of reinforced masonry' - the 'design of prestressed masonry' is discussed in only eight pages (and much of that, in the author's opinion, taken from CP110). The introduction to the draft states that "The draft is based upon current knowledge chosen from reinforced masonry techniques world wide and reinforced concrete technology modified by the experience of the drafters". No mention is made of what the prestressing is based upon.

Part 1 of the code, which has been issued, is one of the few structural codes which contains no mention of such basic design factors as Young's Modulus, the second moment of area, the radius of gyration and deals only with, basically, rectangular sections. (and though section modulus is mentioned this is for free-standing walls only). Part 2 at least gives a value for Young's Modulus - but some engineers doubt its value for brickwork!

It is likely to be at least a year or more before the code is published.

The lack of a code is however unlikely to deter engineers from using the techniques (as evidenced by the papers at this symposium) and it would be a sad day if engineers only did that which the Code suggests could be done.

Though this is not the place to discuss the case that the present trends in codification are more hindrance than help to the engineer, the author would quote just a couple of recent references:-

(i) "Codes..... should embody known and tried principles; their recommendations should be based on actual experience and not on new ideas dreamt up by their authors" (ref 6)

(ii) "The role of codes as an aid in achieving good construction is much in dispute. Should they be somewhere between recent research and standardised proven requirements, telling engineers what is good practice, or should they be precise documents that set out how the task should be handled? Assuming that such precision is even possible, is it in fact desirable, or will innovation and flair be stifled?". (ref 7) Since the author is somewhat anti-code, readers could check the references to see if the quotes are taken out of context.

To counteract the author's anti-code prejudice it should be stated that the draft code in its present form will be of use to engineers and it is probable that the final code will be in a much improved form.

5.3 Research

A surprising amount of research work on reinforced brickwork has already been done and much of the work on reinforced concrete can be 'applied' to masonry. This work can be extended and developed, as always, for we never have 'enough' design data. But to prevent possible delay in application it may be worthwhile recalling that Maillart was building superb bridges in the '20's, Freyssinet did not have the 'advantage' of the accumulation of over 30 years subsequent research and Cadela built daring shell structures long before such research became 'fashionable' (It may be salutory to recall that probably more money was later spent on shell research that on shell construction!).

Since necessity is the mother of invention - and the grandmother of research, it is vital that all researchers (and not just some) should keep in touch with practising engineers - otherwise, as sometimes happened in the past, the research could be sterile, wastefully lavish and of doubtful value.

Sir Alan Harris in a recent excellent article (ref 8) stated that 'Science seeks knowledge for its own sake; engineering is devoted to making useful things'. Engineering has profited from science, but so has science from engineering'.

If the few narrow-minded scientists abandoned their arrogance to the practical engineer and if more practical engineers overcame their unjustified suspicion of the good academics then more fruitful cooperation and rapid progress will result. If there should be a failure to co-operate then engineers should continue to collect their own data from simple site tests.

To aid a sense of proportion in what might appear as just another bilious diatribe against 'academics' it might be helpful to point out that the author has often had cause to bless the BRE and CACA and is hugely enjoying the efficient and helpful co-operation to UMIST. (Not all the good 'places' have been mentioned - that would be impossible but to have ignored them all would have been churlish).

6.0 Potential

If design engineers were to 'ask not should I do this in steel or concrete but can I use reinforced or prestressed masonry' then there would likely be a rapid growth in ideas and applications of these techniques. For experience has shown that they are simple and well within the capacity of small and relatively unsophisticated contractors. The techniques make for rapid and economic construction and the resulting structures can be attractive and durable.

The projects to be discussed in this symposium show innovation, creativity and some daring - but we have only just begun. We need to develop more efficient structural forms than the unimaginative and inefficient plain rectangle. We shall need to be careful of corrosion of steel etc. Since every engineering advance is accompanied by risk taking then it would be prudent to be cautious.

But there are obvious fields for development, based on experience and not mere crystal-ball gazing. There is a wide open field in the multi-storey framed building, in retaining structures, site prefabrication of post-tensioned structural elements, a breakaway from the two-dimensional plate to develop curved and corrugated structures for barrel vaults and walls. (We are conditioned by the drawing board to think in plane surfaces and our conditioning is reinforced on site by the experience that curved shuttering is more expensive than plane, that the cost of bending steel sections can be exorbitant and though masonry curved sections were used in medieval vaults, fortress towers etc. for the last 100 years the bulk of masonry has been mainly employed in 'plate' walls). The bulk of the projects at this symposium are structural and we have hardly looked into civil engineering sub-structures. This is no wild visionary clap-trap but the voice of a hard-nosed practical engineer - and there are many others better qualified than the author to foresee the future developments - it will be fascinating to hear what they have to say today.

7.0 CONCLUSIONS

We need masonry structures to improve the built environment. In the continuing energy crisis we should make more use of those materials which require relatively low energy -input to manufacture and can form energy-saving structures in use. One of the main raisonsd'être of the engineer is that he can provide economic, simple and durable structures and there is a virtually untapped resource in these materials to provide this, and for the good of his soul he should innovate, create and develop - it is a dull dog who only does what has been done before. If man-kind had not been daring (albeit tempered with prudence) we should still be living in caves!

The structures we shall see today, both in the papers and the discussions, look good, were low-cost and are innovative - would that there were more of them! The author hopes that this symposium is onlythe first of a series on this topic, organised by the Institutions, for engineers, by engineers about engineering the use of masonry.

REFERENCES

(1) 'Reinforced Masonry Design' Schneider and Dickey Prentice-Hall 1980

(2) 'Interim Guide for Reinforced Block' ITN 6 Tovey and Roberts CACA 1980

(3) 'Design of Reinforced Brickwork' Curtin, Shaw, Beck and Bray. BDA 1982

(4) 'Structural Masonry Designers Manual' Curtin, Shaw, Beck and Bray. Granada Publishing Ltd. London. 1982

(5) British Standard Code of Practice for the Structural use of Masonry Draft Part 2 'Reinforced and Prestressed Masonry' B.S.I. April 1981.

(6) 'Progressive Collapse Revisited (the problem of structural failure) T.H.W.Akroyd, The Structural Engineer, Dec. 1981. U of 59A

(7) 'Simple Codes can stifle structural technology' Sunley and Taylor Ibid

(8) 'What we want is pragmatic research' Harris N.C.E 10th Dec. 1981.

2. Reinforced brickwork in the George Armitage Office Block, Robin Hood, Wakefield

R. E. BRADSHAW, MSc, MICE, FIStructE, MConsE, Director, and J. P. DRINKWATER, BTech, Engineer; Bradshaw Buckton & Tonge, Leeds

SYNOPSIS. Reasons are given for using reinforced brickwork beams, stairs, walls and piers in the head office of a progressive brick maker. A description is given of the building including the reinforced brickwork elements and the construction.

INTRODUCTION

1. The building is a two storey office block which is the head office for a progressive brick maker - George Armitage & Sons Limited. The total building cost was £767,000 and work on site began in the autumn of 1979 and was completed in late spring 1981.

2. The client's philosophy is to maximise the 'utilisation of bricks' by being able to offer brickwork for a variety of applications including structure and facing.

3. To this end, the company felt it essential that it should construct buildings for its own use in the various forms of brickwork available in order that it may:-

(i) Have practical experience in the construction of such brickwork, including its economics.

(ii) By means of associated testing programmes obtain a better understanding of the structural behaviour of its own products in various applications.

(iii) Show prospective specifiers the various forms of brickwork in use.

4. The company has employed architects to design a high standard of office accommodation for the Company's use and have also constructed reinforced pocket retaining walls, a reinforced brickwork water tank and diaphragm wall workshops.

The company also offers a structural design service through its retained consultants. The progression to a head office incorporating reinforced and prestressed as well as load bearing brickwork and which deliberately probed the frontiers of current knowledge and use was a logical extension of this philosophy. The project was the subject of a competitive tender and the construction contract awarded to the lowest.

5. The form of building chosen by the architect is quite simple and has overall plan dimensions of 13m by 60m. (Fig. 1). The central area incorporates the entrance, reception and main stairs and contains most of the reinforced brickwork. (Fig. 2).

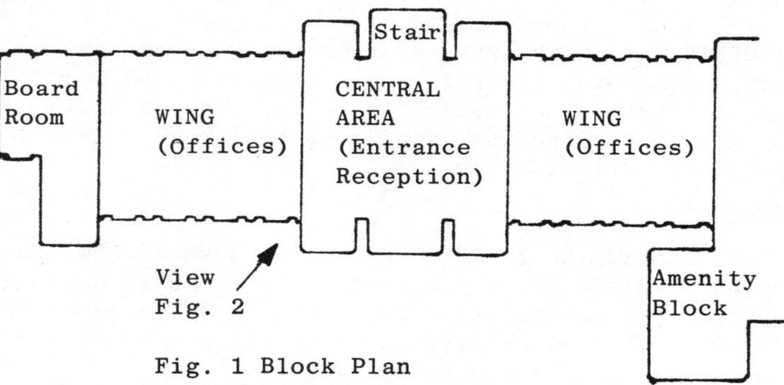

Fig. 1 Block Plan

Fig. 2 View on Central Area Main Entrance.
Post Tensioned Storey Height Wall is above Entrance.

7. The wings to either side have clear spans for the full 10.5m internal depth of the building at both 1st floor and roof level to provide maximum flexibility of office layout.

8. The main support walls to the wings are punctuated with brick piers which provide a shadow line to the brick facade.

STABILITY

9. Is provided by shear walls to either side of the central area and at the ends of each wing.

10. Before work commenced on site, a demonstration structure incorporating reinforced brickwork beams and slabs similar to those proposed for the office block was designed and constructed. Also a series of reinforced brickwork beams were built at the brick factory and tested at Leeds University.

11. The experience and knowledge gained from this confirmed the buildability and structural potential of the proposed design.

THE CENTRAL AREA

12. Incorporates four parallel reinforced brickwork frames two storeys high.

13. Internal spans are 4.2m and 6.7m with 1.5m clear cantilevers at each end except for the two rear cantilevers at roof level over the stair which are 2.7m clear. (Figs. 3 and 4).

14. The design was in accordance with S.P.91 (Ref. 1) as the design and construction of the reinforced brickwork had been completed before the Draft BS 5628 Part 2 (Ref. 2) was issued for comment.

15. The frame was analysed as both fully fixed and assuming continuous beams and pin jointed columns. Reinforcement catered for both conditions. (Figs. 5 and 6).

16. Reinforcement within the 120mm grouted cavity at the beam/column junction was simplified to avoid congestion and aid compaction of the grout.

17. Top and bottom beam reinforcement was continuous at the supports and the column bars stopped at the underside of the beam. Two straight column splice bars were placed centrally.

18. Insofaras possible, within the architectural requirements, bonding patterns were chosen such that the bricks were naturally loaded across the bed faces although strength on end and side was still quite acceptable.

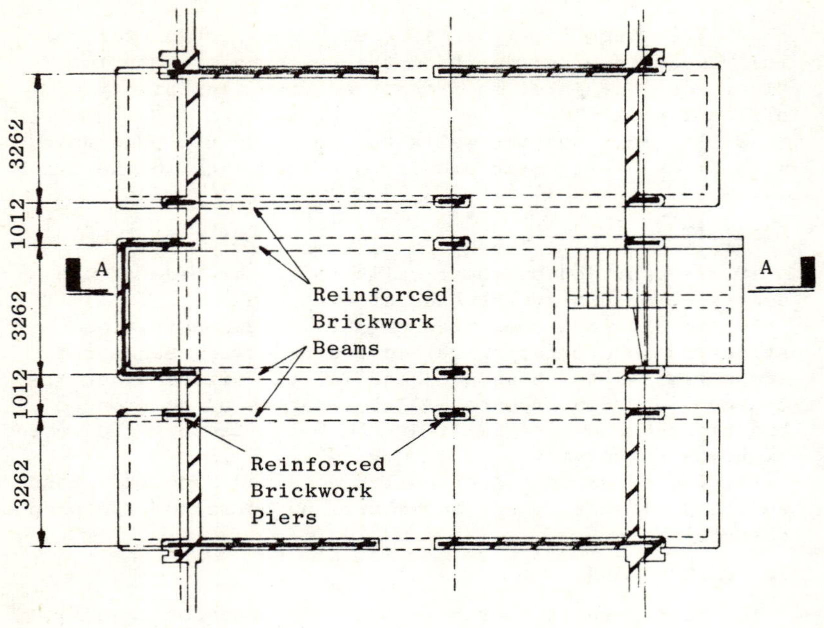

Fig. 3 Central Area. 1st Floor Plan

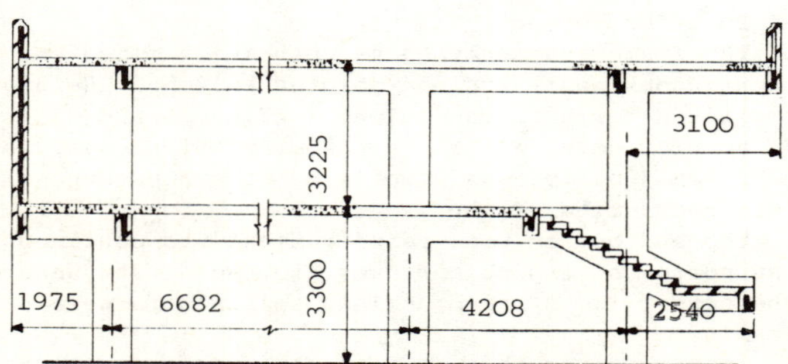

Fig. 4 Central Area. Section A-A showing Reinforced Brickwork Beams and Columns.

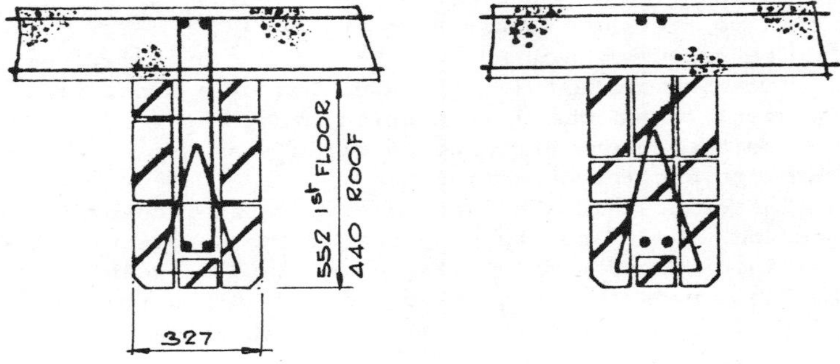

Fig. 5 Typical Beam Cross-section

Fig. 6 Reinforced Brickwork Beam Construction - 1st Stage

19. In certain instances this was not possible and beams formed in stretcher bond were necessary. Tests showed that the beam strength was not impaired since crushing was not the limiting criteria and, as in the columns, a flat link was used to tie the brickwork to the core.

20. The core was grouted up with 1:3:2 grout mix (cement; sand; 10mm aggregate) with the addition of a super plasticising agent (Cormix). This enabled the grout to flow freely around reinforcement and also required less use of the poker vibrator to ensure compaction.

21. The main staircase incorporates reinforced brickwork treads and a half landing composed of two layers of solid bricks laid on bed with reinforcement laid between.

22. The treads and landing are supported by stringers of reinforced grouted cavity construction which cantilevers 1.5m. (Fig. 7).

23. The storey height brickwork above the main entrance was post-tensioned vertically to form a storey height cantilever. Tie steel from the internal beam spans were carried through to the end of the wall at 1st floor and roof level and Macalloy couplings anchored through it. The wall was of grouted cavity design with links bonding the two leaves of brickwork across the core. (Fig. 2)

Fig. 7 Stair Stringer and Columns Under Construction

CENTRAL AREA - CONTINUED

24. Macalloy bars carried in plastic ducts projected through the roof slab into the parapet brickwork above and these were tensioned using a calibrated torque spanner to a tension of 100KN giving pre-compression in the brickwork of 1.7N/mm² ignoring the grout completely. The ducts were grouted up and the anchor nuts and plates suitably protected.

25. Flaking the central area and closing it off from the office wings are grouted cavity shear walls which extend from ground floor to roof level. As in the deep beams, these are simply a tied double brick skin with a grout filling. Reinforcement over openings is incorporated within the cavity.

26. Both the 1st floor and roof construction is reinforced concrete with, at 1st floor, a facing brickwork soffit between the pairs of brickwork beams.

OFFICES

27. The office accommodation flanks the central area and occupies both ground and 1st floor levels. A clear span concrete floor of 10.5m sits on reinforced concrete lintels which in turn are supported by a combination of reinforced piers and simple load bearing brickwork.

28. The majority of reinforced piers, overall size 777mm x 553mm contain a grouted core(123 x 123) containing two reinforcing bars. Shear links were built into every third bed joint as the work proceeded up to roof level. Certain of the piers are prestressed (as an alternative to reinforced) and contain a plastic duct enclosing a Macalloy bar. This is anchored in the foundation ring beam and coupled at 1st floor level with a turnbuckle connector. A concrete cap is provided at parapet level to provide a more even stress distribution and seating for the end plate and locknuts. The bar was tensioned to 90KN giving a precompression in the brickwork of 1N/mm², ignoring the grout completely.

29. The piers are designed to give lateral stability to the office wings by acting as vertical cantilevers above 1st floor level. The 1st floor acts as a horizontal diaphragm transferring loads back to shear walls. The flat roof is of lightweight construction and is assumed to carry vertical loading only, although in practise, it will have some capacity for horizontal diaphragm action. Metsec beams at 3.500m centres support profiled metal decking with a built up felt covering.

MATERIALS AND TESTS

30. Preliminary tests were carried out on all structural materials except steel, before construction began.

31. Stainless steel reinforcement was used in the external faces of those piers and beams exposed to the weather.

32. The strength of the bricks was known to exceed the design requirements and typical strengths are given in Table 1.

Brick	Location	Strength (N/mm²)	
		Design	Actual
Armitage Class B	Reinforced bwk beams except band course and cant brick.	69	88
Armitage Red Rustic	All facing brickwork except band course and beams	56	83
Accrington Class A	Stair treads and landing: cant brick in reinforced brickwork beams.	69	145

TABLE 1 BRICK STRENGTHS

33. The selection of suitable sands for different mortar mixes posed most problems as a uniform coloured mortar was required for all facing work and reasonable workability was required without the use of additives for the reinforced brickwork.

34. A coloured sand to Table 1 of BS 1200 was used for the unreinforced brickwork and for pointing. A sand to Table 2 of BS 1200 was used for the reinforced brickwork.

35. The mortar mixes used and strengths are summarised in Table 2.

OBSERVATIONS

36. The relatively novel use of reinforced and post-tensioned brickwork in the dual role of structural material and quality facing material has been a complete success.

37. The bricklayers adapted enthusiastically to this form of construction.

38. There is clearly scope for wider application of the technique, particularly post-tensioning and grouted cavity brickwork.

39. Reinforced brickwork beams might appear expensive compared with steel or R.C. although the cost of a quality, maintenance free finish material should be added to the

Mortar Mix (Cement-Sand)	Location	Strength (N/mm^2) at 28 Days		
		Min. requirements to SP91		Typical Site
		Preliminary	Works	
1-3*(fair faced work pointed with 1-4½)	All reinforced brickwork in the central area	16	11	20
1-4½ with plasticiser and coloured mortar: pointed as work proceeded	Unreinforced brickwork in wings	6.5	4.5	9
1-4½ with styrene butadiene additive (pointed with 1-4½)	Parapets	6.5	4.5	12
1-3-2 grout (cement-sand -10mm agg.) plus Cormix super plasticising additive (Ref. SP1)	Grouted cavity walls, piers, stair beams reinforced brickwork beams	10	7	30

*Styrene butadiene added to mix for exposed work.

TABLE 2 SUMMARY OF MORTAR AND GROUT MIXES AND STRENGTHS

cost of steel or R.C. for a true comparison.

40. Different mortar mixes are often required in different parts of the structure and this must be related to the sequence of construction and whether to point the brickwork as the work proceeds or later.

41. Experience with the construction of the test beams highlighted the fact that bricklayers are generally less familiar with the interpretation of reinforcement drawings and that thorough supervision is needed.

42. Careful consideration must be given to accommodating D.P.C.s.

43. Grouting of small pockets and congested areas was satisfactorily achieved by the addition of Cormix to the grout.

ACKNOWLEDGEMENTS

44. We acknowledge with thanks the interest and enthusiasm shown for the project by the Board of Directors of George Armitage & Sons Ltd and Stuart Bell RIBA - Head of Technical Services. We also thank Donald Foster RIBA of Structural Clay Products Ltd for his considerable interest and support, particularly with regard to developing the structural form to the Central Area.

Architects: John Brunton & Partners, Bradford

Structural Engineers: Bradshaw Buckton & Tonge

Quantity Surveyors: Rex Procter & Partners, Leeds

Contractors: Ackroyd & Abbott Ltd, Sheffield

Contract Manager - Doug Beech succeeded by Eric Newbold due to ill health.

Site Agent: Peter Gamble

REFERENCES

1. 'Design Guide for Reinforced and Prestressed Clay Brickwork'. Special Publication 91. British Ceramic Research Association 1977.

2. British Standard Code of Practice for the Structural Use of Masonry : Part 2 : Reinforced and Prestressed Masonry, Draft for Public Comment, Document 81/10350, British Standards Institution.

3. Examples of the application of reinforced blockwork

S. ADAMS, FIStructE, MConsE, Partner, and J. E. SAUL, BSc(Eng), MICE, MIStructE, Associate Partner; W. G. Curtin & Partners, Mold

SYNOPSIS. The design of reinforced blockwork is now well established but it will be some time before a wider application of this form of construction is realised. Some applications are described to demonstrate the potential of this unique building form which is used far more abroad than in this country. The examples mentioned are all uses which have been found to be suitable answers to specific problems in the authors own practice.

INTRODUCTION

1. The development of design with reinforced blockwork needs to be accompanied by parallel development in the blocklaying trade. Some countries do have basic well established blocklaying trades quite separate from bricklaying recognizing the different approach required. In this way the new craft techniques are developed allowing further scope for the designer. It is in countries adopting this approach that the most development in reinforced blockwork has been made.

Design and Construction techniques specifically relevant to reinforced blockwork are now being developed and we are on the point of thinking of blockwork as a developed technique in its own right, and not as one borrowing methods from brick experience.

2. The greatest development in this country has been in the field of agricultural building. Farmers have been quick to realise the practical economies of blockwork generally and in exploring the limits of unreinforced blockwork they have readily turned to reinforced blockwork. The Ministry of Agriculture Fisheries and Food have helped considerably in this development both in keeping farmers

up-to-date with the latest developments and in generating discussions among engineers to produce acceptable codes of practice.

There are dangers of abuse with all agricultural work and we had problems in the early days where farmers saw reinforced blockwork as a do-it-yourself job for the winter months. One either has to talk them out of it or design and detail with due allowance.

3. Slowly and probably as a result of agricultural use industrial developers particularly in rural areas have realised the economic use of reinforced blockwork for factory buildings. Even in urban industrial areas it is used increasingly but more in conjunction with the traditional steel portal frame to meet the requirements of the Building Regulations on Structural fire precautions.

4. Architects are more and more turning to blockwork for public buildings and offices because reinforcement gives a monolithic structure obviating the use of a primary structural frame where high flexural stresses are encountered.

5. In domestic building the use of blockwork is extensive but so far reinforced blockwork is almost non-existent. This is probably because domestic building is basically gravity construction with very low flexural stresses. Also the 100mm cavity leaf is not suitable for bar reinforcement. The recent change in the Building Regulations with regard to thermal insulation may introduce a 150mm cavity leaf giving more scope for insitu and prefabricated reinforced block components.

APPLICATIONS OF REINFORCEMENT IN BLOCKWORK

6. Bedjoint reinforcement is mainly used for:-

(a) Controlling shrinkage

(b) Distributing lateral loads

(c) Controlling stresses due to vertical displacement e.g. walls built off flexible floor slabs, and to ensourage composite action between beams and walls.

(d) Shear resistance in reinforced blockwork columns.

7. Horizontal block reinforcement (placed within the depth of the blocks) is mainly used to:

(a) Create beams and lintols

(b) Distribute point loads both vertical and horizontal.

(c) To resist lateral loads in walls spanning

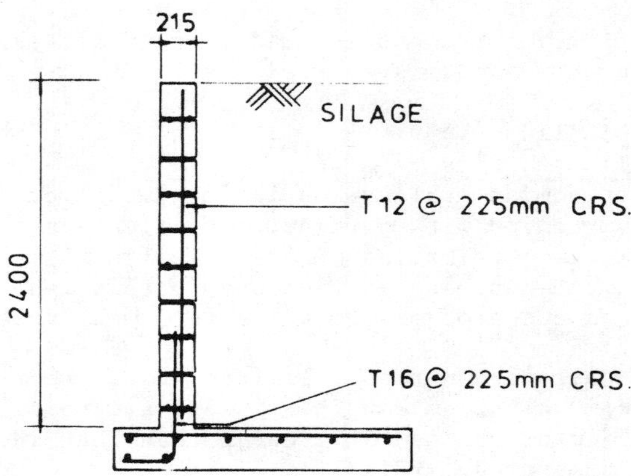

Fig.1. Typical silage wall detail.

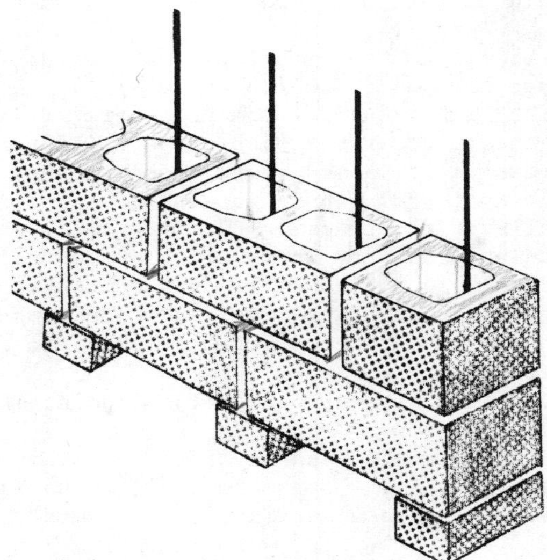

Fig.2. Concrete brick headers used to form clean out openings and to facilitate steel fixing.

horizontally.

8. Vertical reinforcement is used to resist lateral load in walls, piers and columns.

SPECIFIC APPLICATIONS

Retaining Walls and Pits

9. The most common application is in cantilever retaining wall construction with particular application to the storage of silage, grains and other agricultural materials. A lot of research has been carried out in this type of application and codes of practice are very well developed (ref. 1 and 2).

10. These walls are normally limited in height to about 3m and more commonly 2.4m but the flexure due to silage loads in particular can be 70% greater than that due to normal earth pressures. (Fig.1.)

11. A useful recommendation for this type of structure is to start the wall with a row of spaced header bricks to allow cleaning out of mortar droppings at the base prior to concreting. (Ref. 3) This also allows fixing of reinforcement at the base after having first built the block wall and so avoid having to thread the block down the 2.4 metre high rods. (Fig.2.).

Special blocks are now available and suitable mixes developed to ease placing of the concrete fill.

12. Walls with high lateral loads need careful site supervision to ensure accurate location of reinforcement and thorough cleaning of voids prior to concreting.

13. Low head (up to 3m) liquid retaining structures may also be of cantilever construction but with small tanks (e.g. for recirulating slurry above or below ground) walls may be designed to span horizontally or a combination of both. Waterproofing can be provided by a sand and cement render or a separate membrane. Care is needed in controlling shrinkage.

14. This type of construction is also popular in domestic swimming pools.

15. Stock pen walls also usually fall into this category of two way spanning walls. Adequate cover to horizontal bed joint reinforcement is essential in an agressive environment.

Tall Buildings

16. With the development of lighter roof, floor and wall construction unrestrained wall heights of more than 3m present a problem for traditional gravity construction. The problem is not due to high gravity loads, on the contrary

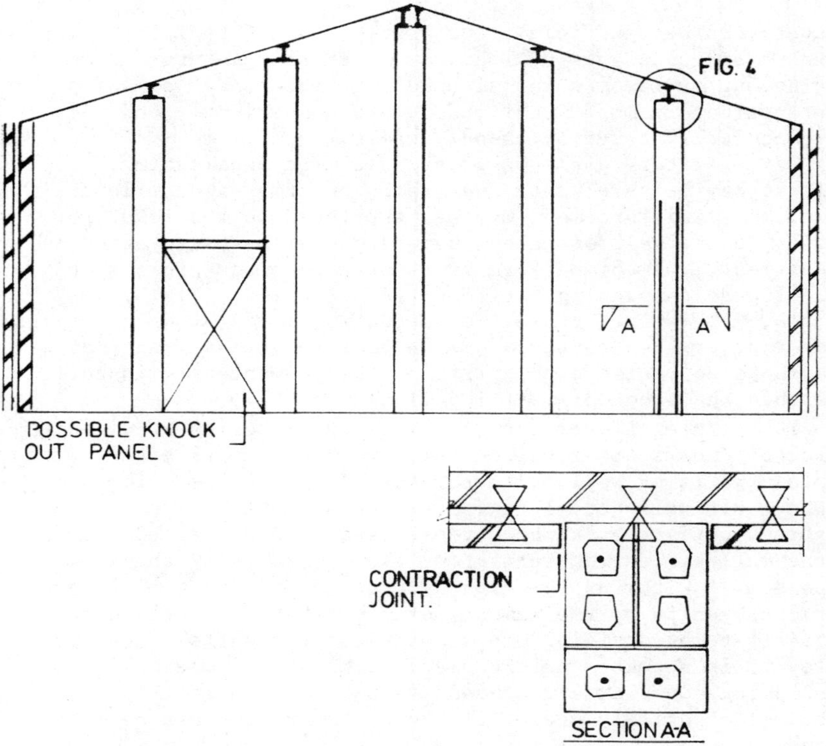

FIG. 4

A A

POSSIBLE KNOCK OUT PANEL

CONTRACTION JOINT.

SECTION A-A

Fig. 3. Factory gable wall with reinforced block piers.

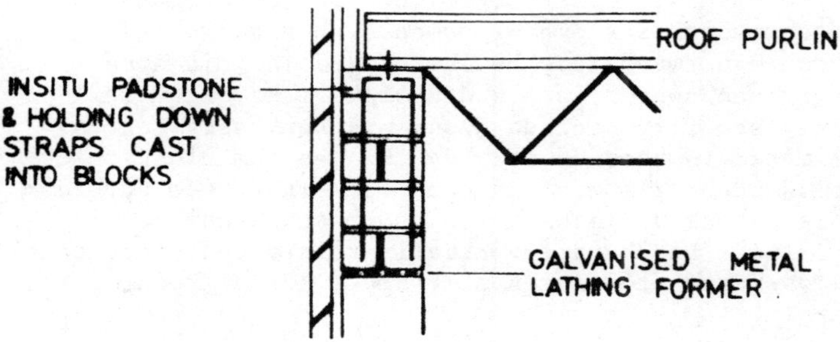

Fig.4. Holding down detail for roof purlins.

these can be beneficial, the problem is due to lateral loads. These are usually wind loads or internal pressures caused by wind but may be mechanical handling loads taken on a separate structure more usually but more conveniently and more economically taken by masonry walls.

17. Factory gable walls are a common application. These may be pure cantilever walls or propped cantilever walls. Reinforcement may be restricted to the lower portion of the wall or taken full height with stages of curtailment. A common form is reinforced block piers with wall panels spanning horizontally. (Ref. 2) Fig.3.

18. This form of wall construction is ideal when purlins or roof trusses correspond with pier locations when roof anchors to resist wind uplift can be conveniently installed within the blockwork. (Fig.4.)

19. Internal compartment walls in factories or dividing factory units are ideal in reinforced blockwork either with plain walls or walls with reinforced piers similar to the gable wall detail Fig. 3. One advantage with this construction is that knock out panels can be concealed within the wall by casting reinforced block lintols in the wall panels. With further expansion or the use of two or more factory units by the same manufacturer these panels can be easily removed giving free access between units. Conversley it is a technique the practice has found useful in breaking down large redundant factories into smaller lettings, especially where fire compartmentation has become necessary.

20. When large lateral mechanical handling loads are involved reinforced block diaphragm walls may be appropriate (Fig.5). Unreinroced walls of this type have considerable resistance to lateral loads but may become unreasonably large to accommodate large base moments when unpropped. Reinforcement can reasonably be introduced in this area and still give an appropriate width of wall at crane rail level. The use of reinforced blockwork for heavy lateral loads is limited in this country by the maximum block thickness which is about 225mm. In countries where the development is more advanced blocks up to 600mm thick can be obtained. These blocks are virtually reinforced concrete column formwork and procedures similar to normal reinforced concrete apply.

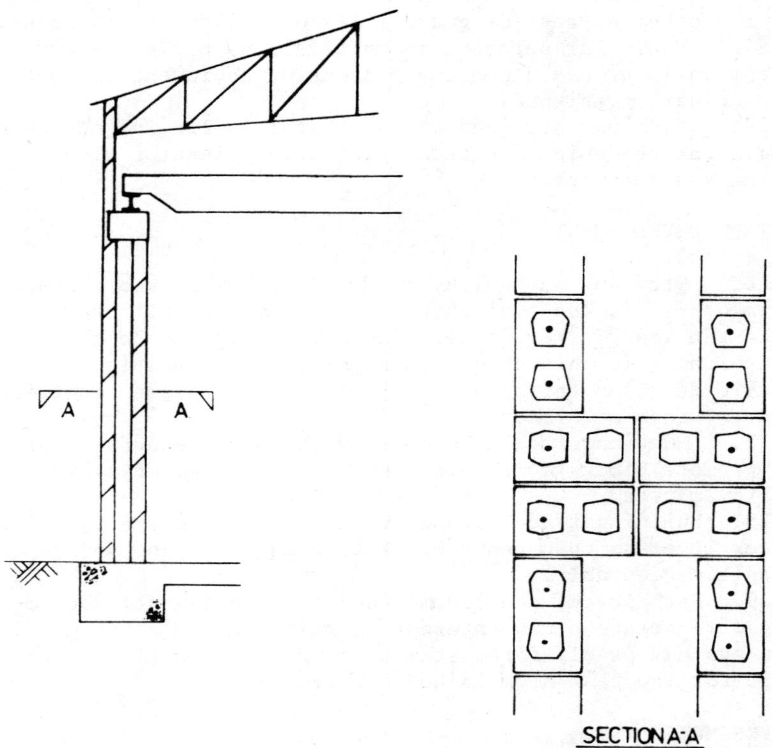

Fig.5. Gantry crane wall using reinforced block diaphragm walls.

Architectural Reinforced Blockwork

21. This is used to a limited extent wherever fairfaced blockwork is required without displaying a structural frame work in other materials. Beams and lintols can be formed within a wall with no disruption of surface finish. Many architects appear to be restricted by the traditional frame work and do not explore the unique opportunity of insitu cantilever masonry to produce more interesting structural forms.

22. However, there are numerous traditional forms which can be more conveniently constructed using reinforced blockwork.

23. Gallery parapets subject to crowd loading are well suited to cantilever blockwork. The same principle applies

to any other parapet or guard rails on stairs and landings.

24. Vehicular parapets in multi-storey car parks and blast walls around dangerous processing equipment can be conveniently designed.

25. High unrestrained walls in stair wells and entrance halls can be designed without projecting piers or other primary structures.

FUTURE DEVELOPMENT

26. With such a healthy supply of cheap brickwork and formwork it is doubtful whether this country will ever take the lead in reinforced blockwork. It may be that Consultants in this country designing structures in the Middle East and third world will learn from these countries and try them at home.

27. Much more could be done by the block manufacturers to promote blocks more suitable for reinforcement as in North America.

28. Building trades should look more to the specialist block layer so that heavy block techniques may be more readily acceptable.

29. Perhaps the new 150mm inner leaf in domestic building will create a new interest in reinforced block components and panels for system housing but we will have to wait for the next boom in house building.

REFERENCES

1. The ultimate strength of reinforced infilled concrete block columns. W.J. Holmes - W.G. Curtin and Partners.

2. Interim design guide for reinforced concrete blockwork subject to lateral loading only. A.K. Tovey and J.J. Roberts. Cement and Concrete Association.

3. Design in blockwork. M. Cage and T. Kirkbride.

4. Reinforced masonry cantilever construction

R. J. M. SUTHERLAND, BA, FICE, FIStructE, FIHE, MConsE,
Partner, Harris & Sutherland, London

SYNOPSIS. The crucial questions are when to reinforce (or prestress) a masonry cantilever, and when to convert a wall conceived as reinforced concrete into reinforced (or prestressed) masonry. The advantages and limitations of reinforced brickwork and blockwork are considered in relation to four case-studies, followed by some thoughts on prestressing and the danger of corrosion with a look at the potential advantages of the new code BS 5628: Part 2, issued in draft for comments over a year ago.

INTRODUCTION

1. When faced with the need to retain earth or to build a high boundary wall most engineers think immediately of reinforced concrete. If there is a need for a visual match to adjacent masonry construction or, if the appearance of concrete is ruled out, they face the concrete in non-structural masonry - either brick or block as appropriate. Very seldom do they think of reinforcing or prestressing the masonry itself.

2. This reluctance to use reinforced masonry is frequently attributed to the inadequacy of the present code CPIII, the relevant clauses of which compare unfavourably with those in American and other codes. For two reasons this explanation is unconvincing.

3. The first reason is that one can achieve a very considerable structural performance with masonry cantilevers working within the admittedly confined recommendations of CPIII. This has been shown in the handbook SCP10 issued by Structural Clay Products (1).

4. Boundary walls up to 5-6 metres high and roughly 275-340 mm thick are possible with central reinforcement and earth retaining ones of 2.5-3.0 metres height of the same

thickness all within the limits of CPIII. Further with "pocket type" construction cantilever walls retaining 4-6 metres of earth are both feasible and economical with thicknesses not much greater than with reinforced concrete.

5. Secondly the use of reinforced masonry in USA although sometimes quite impressive is only regional and sporadic in spite of more liberal permissible stresses of long standing.

6. Stress levels are not the main problem and nor it seems is cost. Ignorance is one difficulty and lack of faith follows from ignorance. However there are some undoubted practical restrictions, which will be discussed below in relation to actual structures. It should seldom be difficult to live with these restrictions given that they are recognised early enough. The real question is when would it be advantageous to use reinforced masonry for cantilever walls. Conditions will vary but the best general answer must be "more often than you think".

7. In the following section four examples of reinforced masonry are reviewed, all based on personal experience and each chosen as representing a different type of cantilever wall. All four walls, or groups of walls, were built with bricks but, with slight changes, concrete blocks could have been used and, in three cases at least, the masonry was not stressed critically even even in relation to the modest limits in CPIII. These walls have been chosen partly to illustrate points which will be discussed at the end of the paper and partly because having been completed for up to nearly twenty years they can now be observed in at least semi-maturity.

CASE STUDIES: REINFORCED MASONRY

(a) Retaining walls on Ethelred Street Housing Scheme, London

8. In this housing scheme, largely built in loadbearing brickwork, split-level planning allowed garages to be slotted in below several of the housing blocks with partially sunk roads leading to them and pedestrian ramps up to raised courtyards. This entailed the use of considerable lengths of retaining wall mainly of heights up to only about 1.5m but subject to the surcharge of vehicle loading

9. Figure 1 shows details of these walls which are clearly beyond the gravity range for their thickness.

10. The compressive strength needed was well within the capacity of the facing brick used throughout the project,

all shuttering was avoided and the reinforcement was easily carried up through the parapets to ensure good resistance to possible vehicle impact. Curves could be negotiated with ease, as could variations in differential levels at ramps. The quantity of reinforcement needed was modest and there was the added advantage of being able to use similar details in the balcony parapets of the flats.

11. This application of reinforced brickwork has been given prominence because, although structurally it is far from exciting, it emphasises the convenience and economy of the technique especially where, as here, it can be repeated on quite a large scale.

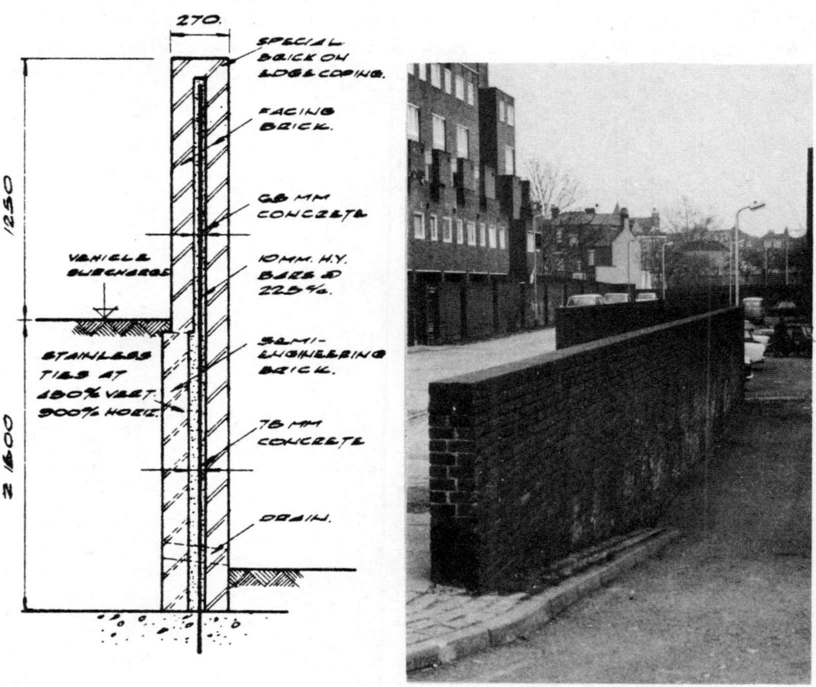

Fig.1 simple brick retaining walls at Ethelred Street

12. The work was completed in the late 1960s and if it was to be done again today the only improvements which spring to mind would be to apply some waterproofing at the back of the walls and possibly to use sulphate-resisting cement. There are slight signs of efflorescene on the faces of the retaining walls but not on the parapets which could mean that some water is passing through which in the long-term might reduce the life of the reinforcement. There are no signs of rust or disruption at present.

(b) Prefabricated boundary wall panels

13. With concrete blockwork, boundary walls as thin as 140mm can readily be stabilised by adding vertical reinforcement but, using standard bricks, such walls need to be about 270mm for grouted cavity, 300mm for brick-on-edge Quetta bond or 337mm for full Quetta bond. At these thicknesses the walls tend to be just about stable as gravity structures for most heights.

14. As an experiment in prefabrication and in making stable boundary walls with the minimum of materials the arrangement shown in Fig.2 was tried out in collaboration with Butterley Building Materials Ltd.at one of their works.

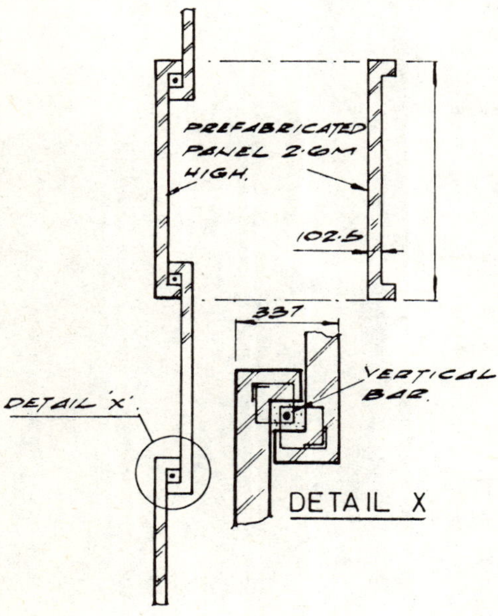

Fig.2 Prefrabricated reinforced brick boundary walls

15. There was little difficulty in handling the panels, but on site care would need to be taken to ensure neat joints between the panels with the brick courses lining up. Using the same section without prefabrication the bay widths and wall heights could be increased quite markedly, still giving great stability with a minimum of material.

(c) Essex University Student Tower Blocks

16. Another quite different type of reinforced brick cantilever, this time in Quetta bond, was used in the early 1960s for first two of six 14-15 storey brick tower blocks at Essex University. The structural form of these buildings was thought of from the beginning as wholly "gravity" brickwork with the wind forces resisted by an assembly of individual but laterally connected vertical cantilevers, the 100mm floor slabs being considered too flexible for the effective transfer of bending moments. (See Fig.3 below)

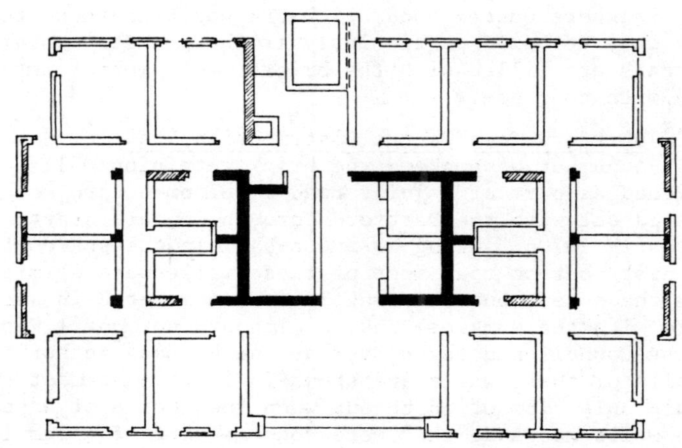

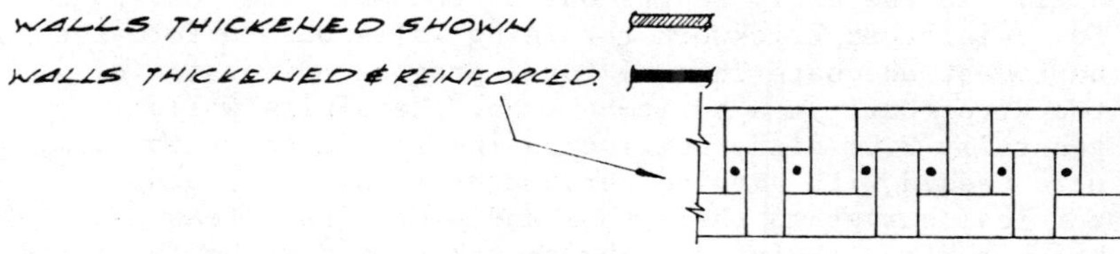

Fig.3: walls reinforced to resist direct tension

17. On this basis the wind forces were seen as distributed between the wall units, whether T, L or I shaped in proportion to their stiffnesses. The flanges of these units thus had to be designed not for bending perpendicular to their faces but for direct tension counteracted by the compression from the dead weight of the structure above or for extra compression added to the full direct stress.

18. In extreme wind the flanges of some of the stiffer wall units were found to be quite seriously in tension - at least theoretically - and as a safeguard these walls were increased in thickness to 337mm for the lower 7 storeys of which in the end only the bottom four were reinforced with vertical bars. A more sophisticated analysis used on the later towers showed that this steel could be dispensed with (2) (3).

19. This example clearly shows that it is often sensible to reinforce masonry locally where there is a need rather than to apply a minimum percentage of steel to all of it or to abandon it completely for reinforced concrete. Here and elsewhere Quetta bond brickwork has been found to be very easy to build particularly if the pockets containing the bars are filled with the bricklayer's mortar rather than with concrete.

(d) Dartford Town Centre Distribution Road

20. A number of pocket-type brick retaining walls were designed as part of a joint road development project carried out with the Dartford Borough Council starting in the early 1970s. Owing to a cut-back in the scale of the new distribution road most of these walls were eliminated from the scheme but two lengths were completed in accordance with the original design although not until Kent County Council had taken over as the highway authority. Details of these walls are shown in fig.4. At most they retain only 2.5m of earth but when the choice of a pocket-type walls was made they were seen as part of a family of walls of up to about twice this height.

21. In the early stages one of the main uses envisaged for reinforced brickwork retaining walls was to form the narrowest adequate channel for a proposed diversion of the River Dart in a built-up area. Retaining walls generally 2.0m high, but increasing in places up to 4.5m, were needed with varying curvatures in plan, smooth vertical curves to their tops and, for visual reasons, a brick surface. Reinforced brickwork seemed to be the ideal choice for such conditions and the pocket type was the only one suitable for the greater heights.

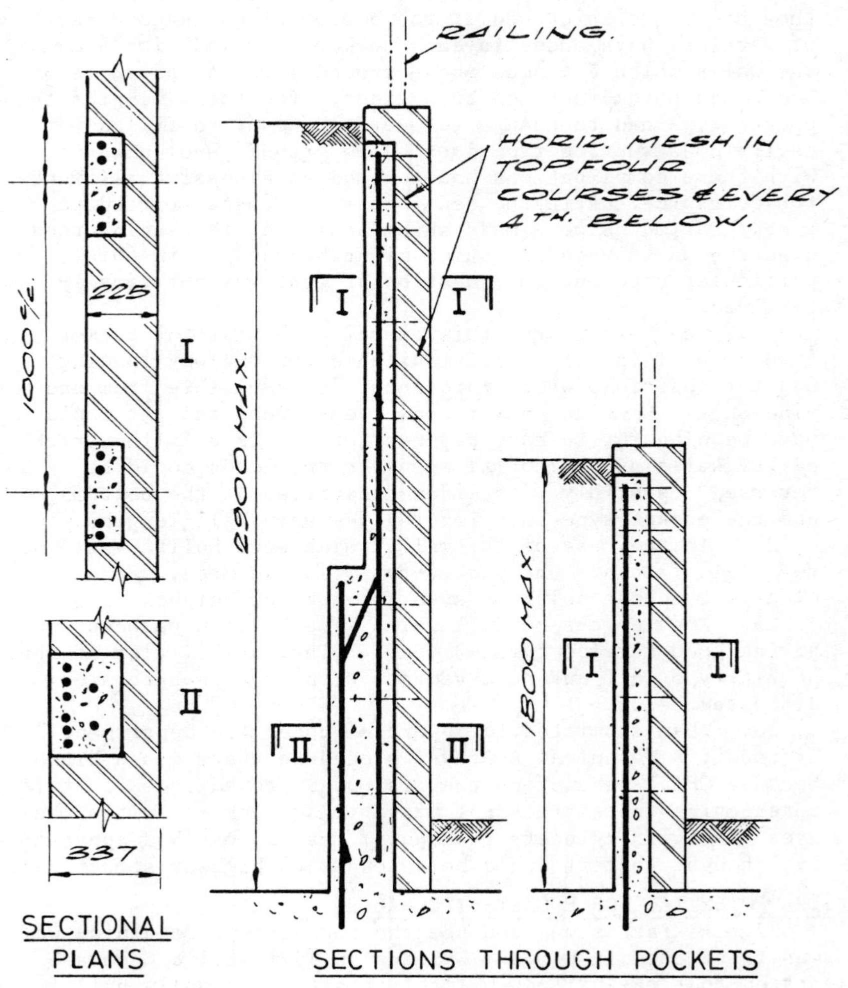

Fig.4. Sections of pocket-type retaining walls at Dartford

22. There were several design complications. The river walls had to resist at least a limited reversal of load with the river in flood and no filling behind; for this they had to be thickened at the bottom with a second set of vertical bars added in each pocket. In addition some of the walls which extended above ground level as parapets had to be brick-faced on both sides; for these lengths the pocket type had to change to Quetta bond or to filled-cavity construction from just below ground level upwards. With these complications and perhaps an excessive allowance for reinforcement in the bed joints the costs started to rise. Estimates were made which indicated that reinforced concrete walls faced in brick might be cheaper in this particular case but the question of cost was not finally resolved.

23. The lesson from this exercise is that the pocket type of wall is only really suitable for one-way loading and for positions where throughout it is visible from one side only. Looking back it now seems clear that it would have been better to have planned for concrete filled brick cavity walls as the normal whenever the loads could be reversed - possibly with widened cavities at the bottoms - and the pocket-type only for the few very tall lengths.

24. In the case of the walls which were built - with no reversal of load - the pocket-type, as executed, was clearly the best solution even for the low heights of wall.

25. The lengths of wall which were built have been giving good service for 2-3 years. The capabilities of the pocket-type wall and achievements with it elsewhere are discussed below.

26. When submitted for approval under the Department of Transport's Technical Approval procedure there was a hitch because CPIII was not on the list of approved codes. It is interesting that structural masonry had sunk so low in the eyes of civil engineers that until then no one had apparenly thought that it could be relevant to highway structures.

THE POTENTIAL FOR POCKET-TYPE WALLS

27. As far as one can see the pocket-type retaining wall was first used in U.S.A. Writing in 1971 Abel & Cochrane stated that the tallest wall of this type actually built was 7.3m high (4). In Britain the tallest equivalent wall of which details have been published is 3.9m high. It was designed by P Maurenbrecher (5) and completed in the mid 1970s. Deflections were measured over the first 517 days and were less than 25mm although the movement, probably mainly due to the earth below, had not quite stopped after

this time.

28. The Author's ideas on how to prefabricate pocket-type retaining walls in panels are given in another paper (6). Full scale trials carried out on panels 3m high by the Butterley Company showed that such prefabrication was perfectly feasible, probably for walls up to at least 4m height. On a large civil engineering project panels could be assembled at leisure, even under cover, to ensure that the rate of building the walls was compatible with the progress of the earworks.

29. It is understood that R.E.Bradshaw has carried out a number of designs for pocket-type walls in recent months. It would be interesting to hear more about these.

CONCRETE BLOCKWORK

30. In many ways concrete blockwork is easier to reinforce than brickwork. Although reinforced concrete blockwork has been used with success for cantilever walls, especially on farms, it does not seem to have been used as much as reinforced brickwork nor on such an adventurous scale.

PRESTRESSING

31. So far the whole discussion about cantilevering in masonry has been centred on reinforced sections. Why not prestress them?

32. The pros and cons of prestressing masonry are discussed in another current paper (7). The conclusions reached, which may be disputed, are that for solid sections anyway the serviceability requirements for prestressing combined with other largely practical factors are virtually bound to make prestressed walls more expensive than reinforced ones. With ribbed and cellular sections prestressing becomes more efficient but even here costs are likely to be greater than with reinforced walls. However, R E Bradshaw has successfully completed some prestressed cellular (diaphragm) walls 2.5M high to retain grain.

33. Despite probable extra costs there is no doubt that prestressing gives a better performance. For the same thickness and masonry strength prestressing ensures smaller deflections and virtually eliminates both cracks and doubts on shear strength. Prestressing may well be the Rolls-Royce treatment but the better performance must both be needed and evident. Few people mind whether their meat is delivered in a Ford van or a Rolls, and even fewer would be willing to pay extra for the latter. Increased shear strength - the trump card of prestressing - is being shown

increasingly not to be a problem with cantilevers (8).

34. Partial prestressing of masonry walls could prove more attractive all round than the full "no tension" treatment but still the benefits need to be seen.

NEW CODE BS 5628: PART 2 (DRAFT)

35. Figure 5 below compares the recommended bending strengths (based on bricks) in the new code with present British and American figures.

36. Whatever the final basis is for calculating resistance movements in accordance with BS 5628: Part 2, it is likely to be between the two sets of figures given in Fig.5 This will give a very substantial increase over the CPIII performance level especially for the higher unit strengths, well into if not beyond the American tabulated range. However it is worth remembering that with elastic (allowable stress) analysis a theoretical increase of over 50% of the figures given is often possible by reducing the steel stress. Also with retaining walls where the bending moment is proportional to the cube of the depth of fill the increased height made possible will be much less than these increases in strength may indicate at first sight.

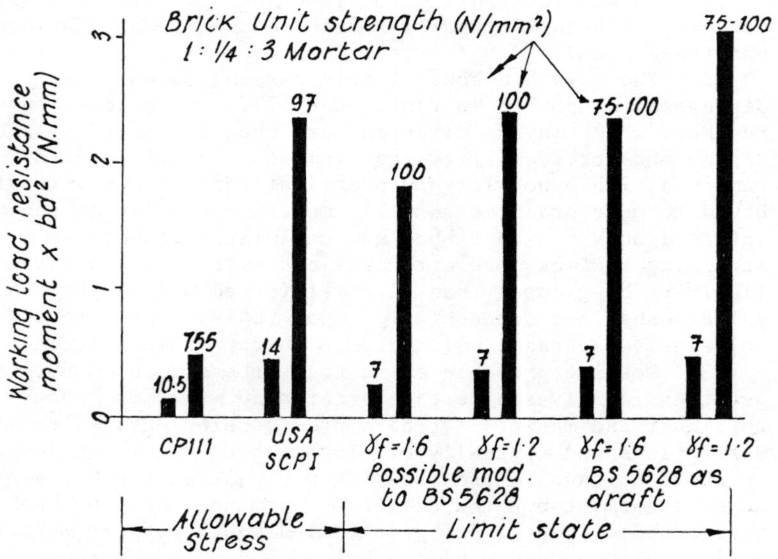

Figure 5: Comparison of "working load" resistance movements (based on bricks)

CORROSION

37. The great unresolved question is what precautions need to be taken to ensure a long life for reinforced masonry especially in very exposed conditions as in retaining or boundary walls. How does reinforced masonry compare with reinforced concrete? Can bricks and mortar or blocks and mortar be considered as part of the cover?

38. There is strong evidence that in time carbonation will extend at least through a half brick skin or the equivalent. (10)(11). The rate of penetration is uncertain but appears to be greatest with weak mortar and porous masonry units.

39. In spite of the likelihood that carbonation will penetrate to the steel there is no certainty that severe corrosion must follow. Both moisture and oxygen are needed for rusting and 100-150mm of brick cover to the reinforcement in a Quetta bond wall could well protect the bars from rusting much better than 20mm of cover in a bed joint even though both are fully carbonated. At present factual records of corrosion are not only sparse but conflicting.

40. We need more evidence not just of carbonation but of the long term effect of this on reinforcement in masonry of all types in differently exposed positions. BRE, BCRA and CACA could usefully combine their resources to find and, with permission, cut open typical areas of reinforced masonry structures 10 or preferably more than 20 years old to check both on carbonation and corrosion.

41. Until we know more about corrosion, designers would do well to use austenitic stainless steel reinforcement (or the new stainless clad bars) in exposed situations. However there seems little need to use stainless steel in pocket type walls or in filled cavities, as long as cover of dense concrete is provided around the bars in accordance with good practice for reinforced concrete which has been proved over many decades.

42. The cost of supplying stainless coated bars varies from 4 to 6 times that of equivalent high-yield steel but spread over the whole cost of a project the extra is likely to be small.

43. Having accepted stainless-coated or pure stainless bars, cover needs only to be that needed for bond and the ways of incorporating steel with safety increases enormously. Although some confirmatory tests would be reassuring there is good reason to think that the thickness of mortar cover needed for bond,where the bars are within or sandwiched between units, is very small.

CONCLUSIONS

44. There are considerable advantages - visual, practical and economic - in reinforced masonry cantilever walls compared with reinforced concrete ones. Corrosion of ordinary HY bars is mainly a problem with thin walls but given effectively corrosion-free reinforcement this problem vanishes completely. As a result the incentive to manyfacture and use special bricks (or blocks) suitable for vertical reinforcement in the most slender walls, becomes very real. The saving in masonry could more than counteract the extra cost of stainless or stainless clad reinforcement. Fig.6 below shows one possible form of slender reinforced wall.

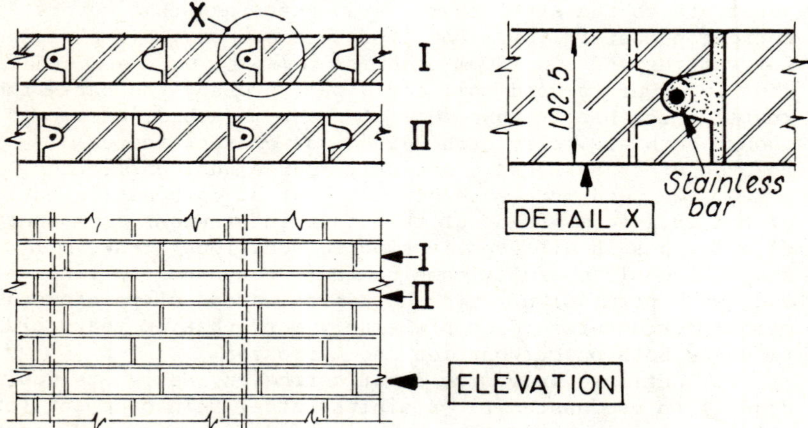

Fig.6 vertical reinforcement in thin brick wall.

REFERENCES

1. Publication SCP10 by Struct.Clay Prod.Ltd: Oct.1976
2. Sutherland R J M Proc. Brit.Ceram.Soc: July 1968
3. Soane A K M & Henry A.W: -do- -do-
4. Abel C R & Cochran M R Proc. SIBMAC, Stoke: 1971
5. Publication SCP15 by Struct.Clay prod. Jan 1977
6. Sutherland R J M Proc I C E Part 1: Feb 1981
7. Sutherland R J M: Paper for 6th IBMAC,Rome: May 1982
8. Edgell G J et al: Paper for 6th IBMAC, Rome: May 1982
10. Kropp J & Hilsdorf H K: 5th IBMAC, Washington: 1979
11. de Vekey R: Symposium on reinf. & prestressed masonry ISE/BCS/CACA London: July 1981

5. Grain silos in reinforced brickwork

R. BEARD, Chief Technical and Research Officer, London Brick PLC, Bedford

SYNOPSIS. The need for storage of grain on farms has grown since the introduction of the combine harvester. Two types of reinforced brickwork silos, one circular and deep the other rectangular and shallow, built by a small but competent builder without specialised equipment or knowledge are described. Examples of the silos built more than 18 years ago have performed satisfactorily, demonstrating the effectiveness of relatively simple design requirements and construction techniques required for reinforced brickwork.

INTRODUCTION

1. Farming of land held in reserve for future brickmaking forms a significant part of London Bricks group activities and has done so for a number of years. The farming is mixed with a large part given over to corn. The introduction of combine harvesters in the 1950's brought with it the need for bulk storage of grain on the farms.

2. In 1955 the Company designed and built five reinforced circular brick silos on one of its farms. The silos were equipped with overhead and floor level conveyors to facilitate automatic filling and emptying via a drying unit for long periods of storage of the grain. In subsequent years similar silos were built on other farms.

3. In 1964 a need arose for a general purpose building used for storage of farm machinery in the winter and bulk storage of grain for short periods following harvest. To meet this requirement the perimeter of a steel frame building was enclosed with a rectangular reinforced quetta bond wall.

4. An important requirement for both types of silo was a simple form of construction which a small but competent

builder could undertake without the need for special equipment. Brickwork is well suited to this concept and the use of reinforcement did not introduce any major difficulties.

5. Both forms of reinforced brickwork grain storage silos, which have been reported elsewhere (1), have been in continuous use and perform satisfactorily. This note describes the more important details about their design and construction.

DEEP OR SHALLOW SILOS

6. The stored grain exerts both the expected lateral force and a vertical force on the retaining brickwork. The estimation of these forces takes a different form depending upon whether the silo is considered deep or shallow. In a deep silo the grain is able to develop arching action whilst in a shallow silo this is not possible - Figure 1.

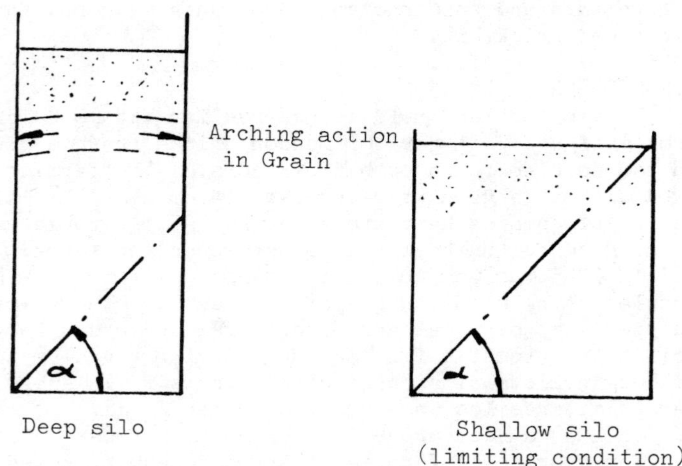

$\alpha = 45 + \frac{1}{2}$ angle of internal friction of grain

Figure 1 - 'Definitions' of deep and shallow silos

7. In these case studies the configuration of the circular silo conformed to a deep silo and the rectangular silo to a shallow one.

CIRCULAR SILOS

(a) General basis of design

8. For deep silos the lateral pressure was estimated from 'Janssen's' equation given in Ketchum (2), which is of the form

$$p = \frac{wR}{u'}\left(1 - e^{-a}\right)$$

where $a = \dfrac{ku'h}{R}$

p = lateral pressure
w = bulk density of the grain (wheat - 800 kg/m^3)
R = hydraulic radius of silo = $\dfrac{\text{area}}{\text{circumference}}$
u'= coefficient of friction between grain and walls of silo (0.4)
e = Naperian log base
k = ratio of horizontal pressure to vertical pressure (0.6)
h = depth of grain

9. Sufficient steel was provided to resist the total tensile force developed by the lateral pressure, the quantity being varied over the height of the silos and accommodated within the brickwork.

10. Because of the lateral pressure and the friction between the grain and the brickwork part of the weight of the grain is transferred to the walls. This can be significant; it is necessary to estimate it and allow for it in the design of the brickwork. The total vertical force transferred to the walls was obtained from

$$P' = \sum Cpu'$$

where P' = total weight of grain supported by wall
C = internal circumference of silo
p = lateral pressure on successive increments of silo height
u' = coefficient of friction between grain and walls of silo

11. In the initial scheme 200 tons of grain were accommodated in a row of five silos each 4m diameter and 5m high

built inside an existing Dutch barn for protection against the weather. This configuration proved very convenient and was adopted for subsequent schemes.

12. A half brick thick wall (105mm) proved quite adequate to accommodate the hoop reinforcement of 7 SWG cold drawn mild steel wire and to support the weight of the grain transferred to the walls by friction, which amounted to about half of the 40 tons capacity of the silo. The principal features of the design are shown in Figure 2.

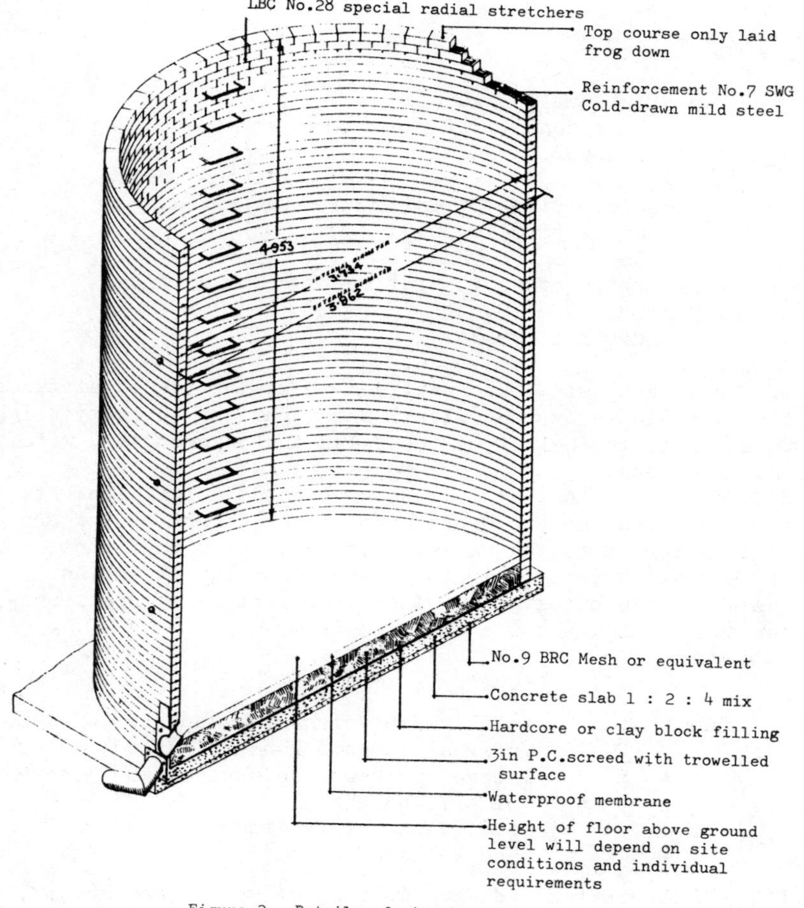

Figure 2 - Details of circular silo

13. To obtain a smooth curved surface use was made of the Company's Fletton radial bricks manufactured for industrial chimneys and curved brickwork features. A special groove was formed in the frogged face of the brick during the brick manufacture to accommodate the steel reinforcement - see Figure 3. The steel was therefore embedded within the bricks and held in position by the bedding mortar which was a gauged 1 : $\frac{1}{4}$: 3 ordinary Portland cement: lime: sand mix. The compressive strength of the bricks was 20 N/mm^2.

(b) Construction

14. No difficulties were experienced with the construction. The centre of each of the five silos was carefully located on the prepared base slab. A scaffold pole was positioned and braced vertically on each centre, the pole being marked in brickwork course heights. A wooden trammel, cut to the radius of the silo and located on the pole, enabled the brickwork to be built to the required diameter and plumb.

15. Reinforcement was cut to lengths allowing 300mm overlap, then laid in the grooves in the bricks and mortared in with the next course of bricks. The grooves proved particularly useful in locating the reinforcement prior to the mortar being laid.

(c) Tests

16. In order to obtain some indication of the validity of the assumptions made in the formula for lateral pressure and to observe if any dynamic lateral pressures were developed during emptying, one silo was equipped to measure the load on the floor and the circumferential strain in the brickwork.

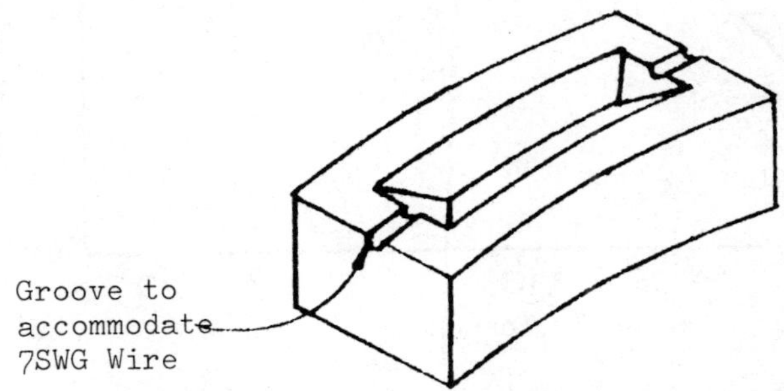

Figure 3 - Radial brick with groove for reinforcement

17. The latter proved singularly unsuccessful, the only indication being that the strains were very small. This implied that they were almost certainly controlled by the brickwork and not by the reinforcement.

18. Measurement of the floor loads, achieved by constructing a temporary false floor supported on three load cells, was more successful. The instrumentation was provided and operated by the Building Research Station. Barley was used for the tests. Floor loads were measured during filling and emptying and compared with the total weight of grain estimated from the volume in the silo and bulk density of the barley. The difference between the two weights was attributed to that carried by the walls of the silo. In Figure 4 the load estimated to have been carried by the walls is compared with the design.

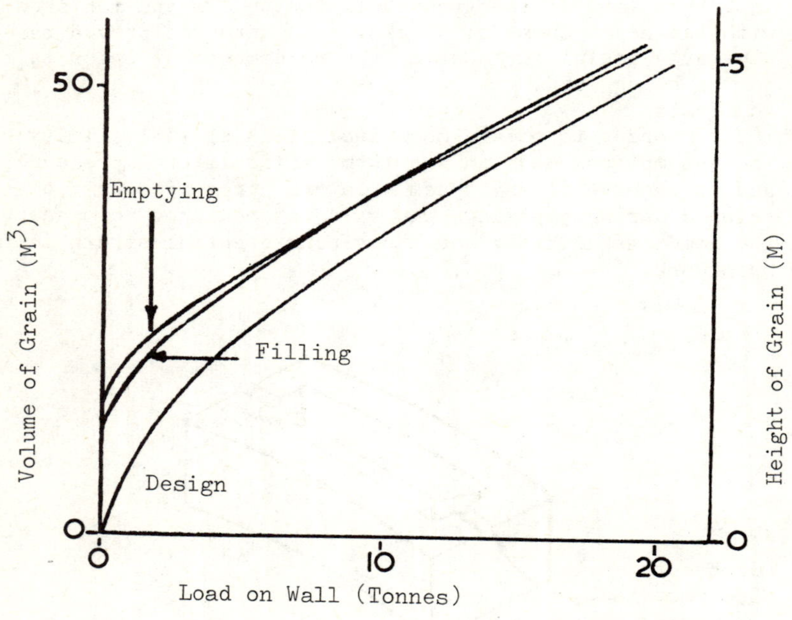

Figure 4 - Wall loads for barley

19. Since the vertical load carried by the silo walls is proportional to the total lateral load exerted by the grain the curves in Figure 4 are proportional to the lateral force. The design assumptions appear broadly correct with perhaps a small overestimation of lateral pressures.

20. No dynamic effects on emptying were observed. All the silos were side emptying and this may have prevented the sudden collapse of the 'arch' set-up within the grain.

RECTANGULAR SILOS

(a) General basis of design

21. The general purpose building was about 24.5m long x 21m wide, with a steel frame supporting an asbestos sheeted roof and carrying a conveyor just below the roof apex. The brick perimeter walls had to contain grain stored to a height of 1.5m and sloping at its angle of repose to the centre of the building beneath the conveyor from which the grain was fed into the building. From this configuration it is quite evident that the design should be treated as a shallow silo.

22. Coulomb's wedge theory was adopted to predict lateral pressures. For the grain sloping at its natural angle of repose the lateral pressure on the vertical surface of the wall is given by the following equation.

$$p = k'wh$$

where
p = lateral pressure
w = bulk density of the grain (wheat $800 kg/m^3$)
h = depth of grain against silo wall
$k' = \cos^2\theta = 0.82$ (for $\theta = 25^\circ$)
θ = angle of repose (wheat = 25°)

23. In addition to retaining the grain the perimeter walls had to keep the stored material dry. Out of a number of schemes considered the most economical and simplest to construct in brickwork appeared to be a cavity wall construction of an outer leaf (105mm) for weather protection and an inner leaf of a quetta bond wall (330mm) vertically reinforced as a cantilever wall for retaining the grain. The quetta bond wall was constructed to a height of 1.68m and at that level the inner leaf was reduced to half brick unreinforced wall to the eaves of the building.

24. Quetta bond produces vertical pockets in the centre of the brickwork at about 170mm centres. Since the lateral force would always be applied from one side, the reinforcement was offset to obtain the maximum lever arm. Fletton bricks having a compressive strength of 20 to 27 N/mm^2 were

used with a gauged 1:$\frac{1}{4}$:3 ordinary Portland cement: lime: sand mortar. The British Standard Code of Practice at the time CP111, 1948(3), recommended a permissable compressive stress in flexure of 1.7 N/mm^2. This was more than sufficient to provide the moment of resistance in compression and an under-reinforced design using 10mm dia. mild steel bars in alternate pockets was adopted. Figure 5 shows the general section of the construction and a plan of the reinforced quetta bond inner leaf.

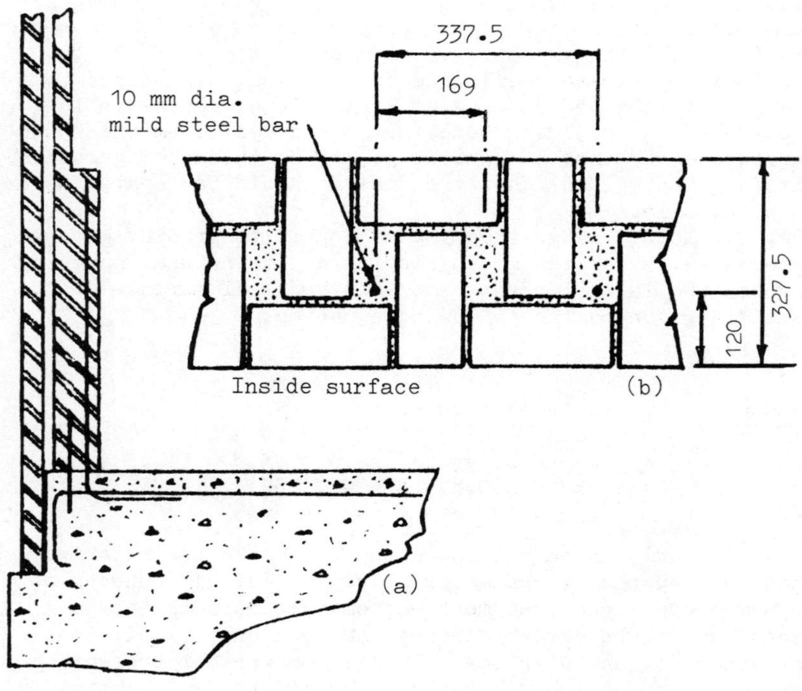

Figure 5 - (a) General design and
(b) detail of quetta bond and reinforcement.

25. No difficulties were experienced with meeting the recommended bond stress between steel and mortar (the bricklaying mortar was used to fill the pockets as the wall was built) nor the shear stress within the brickwork. However, it was decided not to place any reliance on the bond of the first mortar joint and the concrete base and dowel reinforcement was provided across this joint.
26. No horizontal damp proof course was used in the inner leaf. It was considered unnecessary since there was no risk of rising damp through the concrete floor slab and undesirable because it may have reduced the bending resistance and stiffness of the wall acting as a vertical cantilever. It was also decided that because of the reinforced connection between the base of brickwork and the concrete floor slab the brickwork would be restrained against horizontal movement and no movement joints were provided in the walls, the maximum length of which was 24.5m. As far as can be assessed this decision was justified since after 18 years there is no evidence of movement or distress in the masonry.

(b) Construction

27. No exceptional difficulties were experienced with the bricklaying but considerable care was needed to locate the reinforcement. This was important in respect of the width of the wall in order not to reduce the effective depth and thus the moment of resistance and also in respect of the length of the wall to anticipate the position of the vertical pockets in the quetta bond. All main tensile reinforcement and dowel bars were cut to length and located in a pre-drilled timber to hold them in position prior to casting them into the concrete floor slab. Before bricklaying commenced the tops of the main steel were attached to a timber batten which was braced to hold the steel vertical. Bricklaying was then quite simple, the use of reinforcement in alternate pockets allowed the bricklayer to work from one side through the bars.

CONCLUDING COMMENTS

28. One of the important objectives of these applications of reinforced brickwork for grain storage on farms was that the construction should be within the capability of small local but competent builders and this objective was very clearly met. No unusual or special requirements were necessary except to ensure the correct location of the reinforcement - a fundamental requirement in any reinforced masonry.

29. In the examples described the reinforced masonry was protected from the weather to keep the stored grain in a dry condition, it also avoided any special requirements for protecting steel against corrosion.

ACKNOWLEDGMENTS
30. This work is published with the permission of the Directors of London Brick P.L.C.

REFERENCES

(1) DINNIE A. and BEARD R. "Reinforced Brickwork Silos for Grain Storage", Proceedings of the British Ceramic Society No.17, February 1970.

(2) KETCHUM, M.S., "The Design of Walls, Bins and Grain Elevators", New York, McGraw Hill.

(3) BRITISH STANDARDS INSTITUTION The Structural Recommendations for Loadbearing Walls. CP111 : 1948.

6. Design and construction of a reinforced brickwork tank

G. D. JOHNSON, BEng(Tech), MIStructE, Structural Clay Products Ltd, Hertford

SYNOPSIS. A 455cu.m. circular water tank was constructed using high strength bricks. The wall took the form of a grouted cavity structure and reinforced concrete design information was used to obtain the design forces and moments in the structure. This approach was used since there is no detailed information for such works in reinforced brickwork. A summary of the design calculations is given which illustrates the use of the design coefficients used in reinforced concrete designs. Aspects of the design are discussed with particular reference to the choice of permissible stresses used and the introduction of shear reinforcement. The method of construction, waterproofing and the economics of this type of construction are also considered.

INTRODUCTION

Part of an improvement to the installations at the brickmaking factory of George Armitage and Sons included the necessity for a water storage tank. This was required to store at least 455cu.m. to be used in the event of fire danger to the propane storage tanks which fuel the brick-making process. All of the installations were constructed in brick masonry which, apart from their primary functions, forms a focus for demonstration of structural engineering in masonry. The reinforced brickwork water tank was constructed from three hole perforation bricks made at the factory. The reinforcement is incorporated in the brick-work in a grouted cavity which extends from the concrete base to above water level and a protective 225mm wall extends above ground level to form a safety barrier. The size of the tank was intended to be 10.5m diameter with an overall height of 7.25m containing a 5.25m head of water.

Excavations began in April 1979 and, as a result of the discovery of poor ground at the founding depth an additional 1m was excavated. This was utilised to provide an additional 1m depth of water. Thus the overall height became 8.25m with 6.25m of water. This gives a capacity of 541cu.m. The design procedures adopted were based on the principles of reinforced concrete tanks. This is not entirely satisfactory since the assumptions which govern the theory for reinforced concrete water retaining structures do not necessarily apply to reinforced brickwork water retaining structures. However, in order to get the project underway quickly the pragmatic approach of using existing concrete design theory was adopted. Since the tank was designed to withstand the full head of water without backfill, but was eventually to be backfilled, and bearing in mind the large factor of safety inherent in this type of design due to deliberately low steel stresses, it was felt that it was safe to proceed using the existing design information. No detailed design information for construction of reinforced masonry tanks is available in the U.K. The design was based on the recommendations of the British Standard Code of Practice BS.5337(1). The maximum steel stress adopted was 100 N/mm^2 in accordance with BS.5337 for permissible stress designs. This was to try to keep cracks in the structure to a minimum, although it was intended to render the inside walls of the tank to ensure a complete watertight structure. Before rendering the tank was filled and then emptied to allow any crack patterns to develop. The render was then applied, allowed to set, and the tank refilled. This approach is now thought to be over cautious in that the full design strength of the steel could be used thus giving a saving on steel costs. The SBA render should be able to accommodate the larger strains and remain watertight.

DESIGN

Data

Size of tank	:	10.5m dia., 8.25m high, 6.25m of retained water - see Fig. 1
Brick strength	:	69 N/mm^2
Mortar	:	$1:\frac{1}{4}:3$
Steel	:	Characteristic stress 460 N/mm^2 to BS.4449(2)
Concrete	:	Grade 25 to CP.110 : 1972(3)

Design Notation

A_{sv} : Area of shear steel
H : Height of water
H_r : Position of maximum reverse moment
H_t : Position of maximum circumferential tension
K_o : Coefficient for bending at base wall
K_1 : Coefficient for position of maximum circumferential tension
K_2 : Coefficient for maximum circumferential tension
K_3 : Coefficient for shear at base of wall
K_4 : Coefficient for maximum reverse moment in wall
K_5 : Coefficient for position of maximum reverse moment in wall
M_b : Moment at base
M_r : Maximum reverse moment in wall
R : Radius (also a reference symbol for mild steel)
S_b : Shear at base of wall
T : Maximum circumferential tension
Y : High yield steel reference symbol

b : Breadth of section
d : Effective depth
m : Modular ratio
n : Depth of compression zone
p_{bc} : Brick permissible compressive flexural stress
p_{st} : Steel permissible tensile stress
p_{sv} : Permissible shear stress in steel
p_v : Permissible shear stress in brickwork
s_v : Spacing of shear steel
t : Thickness of wall
v : Shear stress
w : Density of water (taken as $10kN/m^3$)
y : Deflection
z : Lever arm

Reference	Calculation	Result
	$H = 6.25m$, $t = 477.5mm$, $R = 5.25m$	
	$\therefore \frac{H}{\sqrt{tR}} = \frac{6.25}{\sqrt{0.4775 \times 5.25}} = 3.95$	
	$K_0 = 0.018$, $K_1 = 0.409$, $K_2 = 0.552$	
Manning	$K_3 = 0.180$, $K_4 = 0.060$, $K_5 = 0.280$	
	Maximum bending at base $= K_0.w.H^3$ $= 0.018x10x(6.25)^3$ $= 43.95$ kN.m	$M_b =$ 43.95 kN.m
	Position of maximum circumferential tension $= K_1H = 0.409x6.25 = 2.56m$ (above base of wall)	$H_t =$ 2.56m
	Value of maximum circumferential tension $= \frac{1}{2}.w.H.D.K_2 = \frac{1}{2}x10x6.25x10.5x0.552$ $= 181.10$ kN/m height of wall	$T =$ 181.10kN/m
	Maximum shear force $= K_3 \cdot w \cdot H^2$ $= 0.180x10x(6.25)^2 = 70.31$ kN/m	$S_b =$ 70.31kN/m
	Maximum reverse moment $= K_4 \cdot w \cdot H \cdot t \cdot R$ $= -0.060x10x6.25x0.4775x5.25$ $= -9.40$ kN.m	$M_r =$ -9.40kN.m
	Position of maximum reverse moment $= K_5H = 0.280x6.25 = 1.75m$	$H_v =$ 1.75m

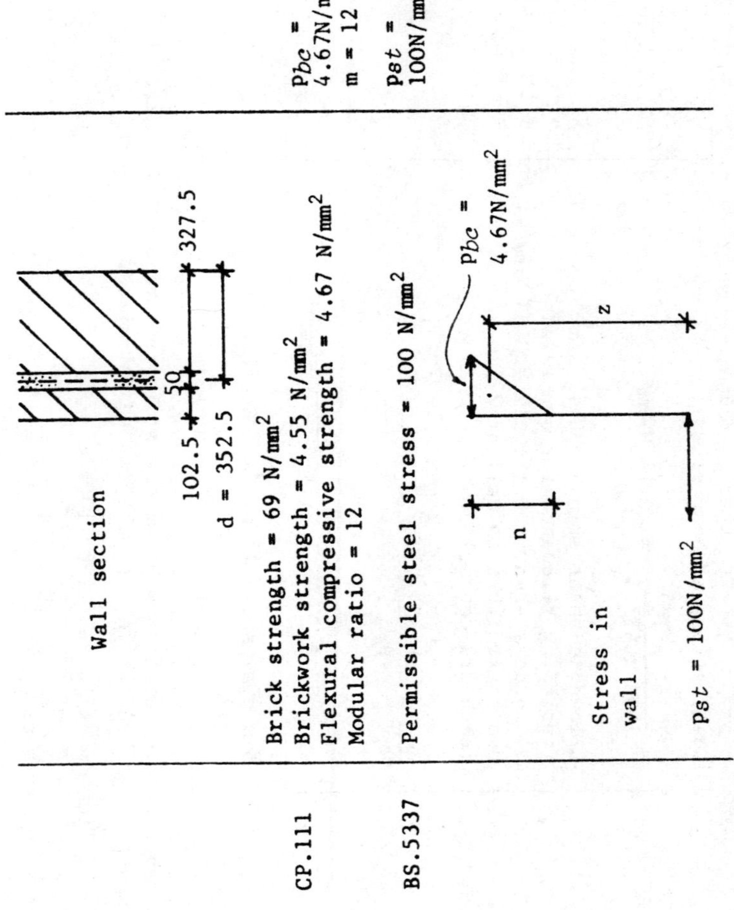
Wall section
102.5
50
327.5
d = 352.5
CP.111
Brick strength = 69 N/mm²
Brickwork strength = 4.55 N/mm²
Flexural compressive strength = 4.67 N/mm²
Modular ratio = 12
Pbc = 4.67N/mm²
m = 12
BS.5337
Permissible steel stress = 100 N/mm²
Pst = 100N/mm²
Stress in wall
n
z
Pbc = 4.67N/mm²
Pst = 100N/mm²

	Shear stress at base of wall	
	$= \dfrac{70.31 \times 10^3}{1000 \times 477.5} = 0.15\ N/mm^2$	$v =$ $0.15\ N/mm^2$
CP.111	Allowable shear stress = $0.12\ N/mm^2$ (including allowance for own weight)	
	$\therefore$ excess shear = $0.15-0.12 = 0.03\ N/mm^2$	
See Fig.1	Use shear links in wall at 225mm c/c (each 3rd course)	
	$\dfrac{A_{sv}}{s_v} = \dfrac{b(v-p_v)}{p_{sv}}$	
	$\therefore A_{sv} = \dfrac{225 \times 100 \times (0.15-0.12)}{115} = 56mm^2/m$	$A_{sv} =$ $56mm^2$
	6mm dia. shear links OK. See Fig.1.	R06 @ 500 c/c

Using the appropriate coefficients the areas of steel required for circumferential tension and reverse bending in the wall can be calculated

tension - 1811 mm^2/m height ≡ Y16 @ 110 c/c
reverse moment - 838 mm^2/m height
- area of steel at base of wall = 1410 mm^2 - ok, reduce to 0.15% at top of wall ≡ Y12 @ 160 c/c

Base design criteria was assuming a hydraulic head of water outside the tank (uplift on base) and followed normal concrete design.

DISCUSSION

Design formula

As can be seen from the summary of the design calculations, the forces and moments are found from the use of coefficients K_0 - K_5. These are based on concrete tank design information. The information is derived from the basic mathematical treatment of tanks and is presented in terms of $H/\sqrt{tR}$ ratio. This is because all tanks of similar materials have the same deflected shape for the same $H/\sqrt{tR}$ ratio. In the analysis of circular tanks in reinforced concrete the assumption is that the modulus of elasticity of the material is the same in tension as in flexural compression and also it is assumed isotropic. This will not be the case with a brickwork structure so that a strict analysis of a reinforced brickwork tank would require a knowledge of the anisotropic properties of the brickwork used and recalculation of the design charts based on this information.

Permissible stresses

Reference was made to British Standard 5337 : 1976(1) for some guidance on the value of permissible stresses to be used to limit cracking in water retaining structures. For concrete structures the stress adopted depends on the degree of exposure of the concrete (cl.4.9 BS 5337).

Class A exposure.	Exposed to a moist or corrosive atmosphere or subjected to alternate wetting and drying.
Class B exposure.	Exposed to continuous or almost continuous contact with liquid.
Class C exposure.	Not exposed to liquid nor to moist or corrosive conditions.

Since there is little experience of masonry used in this type of structure, the lowest permissible stresses for flexural tension, direct tension and shear were used, i.e.-

Direct tension	:	$100N/mm^2$
Flexural tension	:	$100N/mm^2$
Shear	:	$115N/mm^2$

(from Table 4, BS 5337)

As more experience is gained with this type of structure it could well be that these values can be increased with consequent savings in steel costs. The design strength of the H.Y steel is $460 \div 1.15 = 400N/mm^2$ so that steel areas in the walls could be reduced.

Tank wall details
Staggered links
in next course
links
592mm wide
links at
408 c/c
Horizontal
section
Y12 @ 160 c/c
SBA
render
Y16 @ 250 c/c
Y12 @ 160 c/c
6250
Vertical
section
Y12 @ 80 c/c
Y16 @ 110 c/c
6mm dia. links
every 3rd course
575
cork and mastic seal

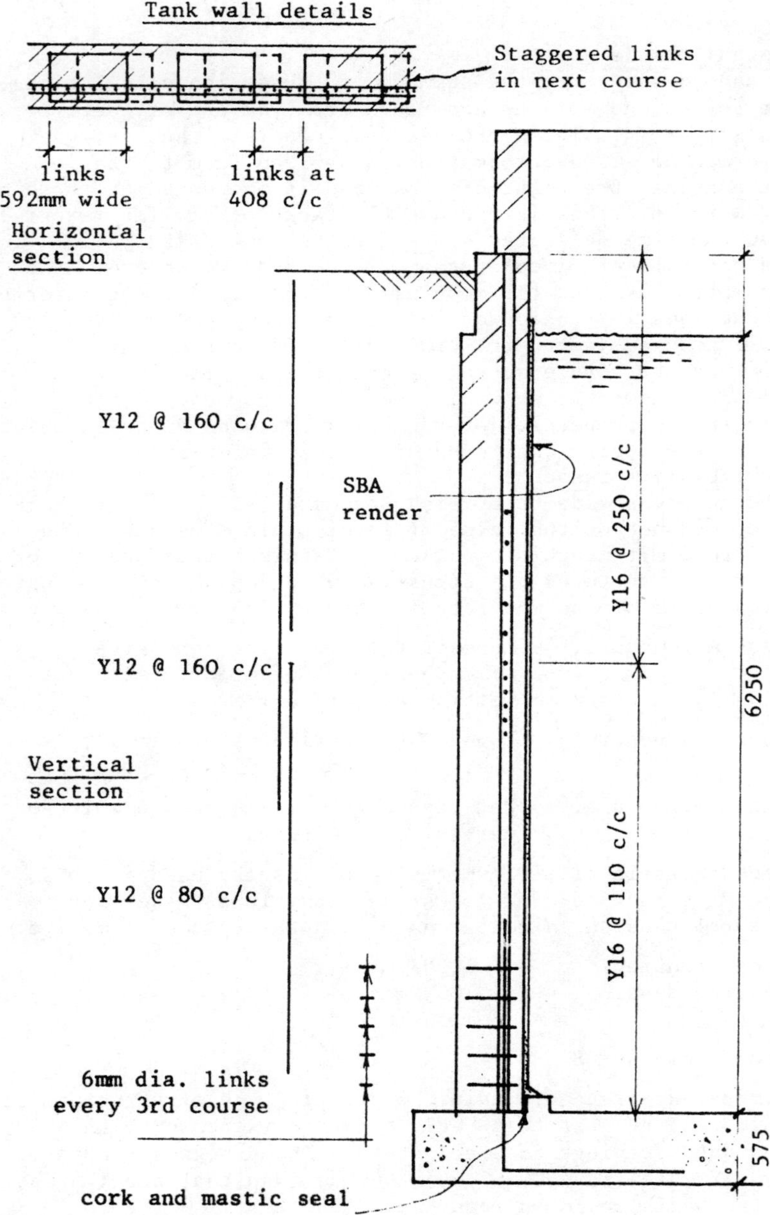

Fig.1.

Shear links

The calculations showed that the permissible shear stress at the base of the wall was exceeded once the depth was increased beyond 5.25m head of water. This could be accommodated either by increasing the thickness of the wall or by incorporating some form of shear reinforcement. Shear reinforcement was used in order to see how this type of reinforcement could be used. To be effective, the shear legs need to cross the width of the wall at right angles to the plane of the wall to intercept any shear cracking that may occur. The method adopted was to use mild steel rectangular loops that spanned the wall cavity. These loops were placed regularly in every third course and staggered on plan. Their behaviour was assessed from normal reinforced brickwork theory, but the actual performance of links used in walls in this way is not well documented. If walls are thickened as required for shear strength then the effects of stepping the walls - i.e. abrupt changes of section - need to be considered.

The method of assessing shear stresses and providing shear reinforcement is open to interpretation depending on the source of reference used. There is no specific guidance for the use of shear links as used in the tank wall. The approach employed here was to calculate shear stresses based on the total cross sectional area of the wall (CP111, cl.321(c) and cl.317). A permissible shear stress of 0.12N/mm^2(permissible shear stress of 0.10 plus an allowance of 0.02N/mm^2 for dead load shear resistance), was taken as the shear contribution of the masonry, the difference between this and the imposed shear stress of 0.15N/mm^2 was assumed to be resisted by the shear reinforcement. The Code of Practice for reinforced concrete which uses permissible stress terms (CP.114 Part 2, 1969)(6) and also BS.5337(1) requires that the shear links should resist all of the shear stress. However, at the base of a masonry cantilever wall it is felt that it is unreasonable to assume that the brickwork shear resistance cannot be considered in view of the small excess shear induced. It is known(7) that the factor of safety for shear in cantilever walls designed to CP.111 can be high (over 9), so the approach used is thought reasonable. Corrosion of the mild steel links is not thought to be a problem since the wall is protected from water by the SBA render and the bitumen paint. In addition the interior of the wall is similarly protected from atmospheric carbon dioxide which is known to promote reinforcement corrosion by carbonation of concrete and mortar thus leading to an acid environment.

Materials

(a) Bricks: The bricks used were "solid" (less than 25% perforations) class "B" engineering units of special quality as defined in British Standard 3921 : 1974[8]. This means that they are deemed frost resistant bricks - whether this is a necessary attribute for a buried water tank, rendered on the inside, would be a matter for opinion. Certainly the bricks above the water line and unrendered, would need to be frost resistant. The manufacturer's quoted strength is 69N/mm^2. This is not an average figure but a lower acceptance figure used for limit state designs. The true average strength of production is around 80N/mm^2 but, since there is an upper limit for flexural compressive strength in CP.111, there was no advantage in using such a figure. In any future designs a laboratory assessment of flexural compressive strength could be used to allow economies in design to be made.

The water absorption was 7.0% and the size of the bricks was the standard U.K. format of 215mm x 102.5mm x 65mm.

(b) Steel: All the reinforcing steel used for flexural tension and direct tension was in the form of high yield deformed bars conforming to British Standard 4449 : 1978[2]. This has a characteristic strength of 460N/mm^2 for all bars up to 16mm diameter. The shear links were of mild steel plain bars, also conforming to BS.4449, with a characteristic strength of 250N/mm^2

(c) Mortar: The mortar used throughout was 1:$\frac{1}{4}$:3 Ordinary Portland cement : hydrated lime : sand mix by volume.

(d) Grout: The grout mix was 1 part cement to 3 parts sand to 2 parts pea gravel by volume. The pea gravel had a maximum particle size of 9mm. The grout slump was specified as 175mm.

(e) Concrete: Grade 25 concrete to CP.110 : Part 1:1972[3] was used for the construction of the tank base. The concrete was manufactured off the site and the whole of the base was delivered and placed in one operation. Grade 25 concrete should have a characteristic compressive strength of 25N/mm^2 when tested at 28 days.

(f) Render: Cement/Styrene Butadine render, 16mm thick in two coats.

Construction

Excavation of the site proceeded using conventional plant. The discovery of unsuitable ground at the intended foundation level called for a new design to cater for the additional 1m depth found necessary. After levelling the base with a cement/sand blinding the base steelwork was fixed, including wall starter bars, and base formwork erected. The base was originally designed as a circular structure but, on the advice of the contractor, it was detailed as octangular to allow some considerable saving in base formwork costs. The shuttering for the concrete nib at the wall/base junction was also set up at the same time, being supported off the main base formwork. The change to an octangular base made little difference to the design calculations. At the centre of the base was cast a scaffolding pole which formed a centre for the construction of the curved brickwork wall. This was later removed and the hole (approximately half the base depth) was filled in with concrete. The progress of work to this stage can be seen in fig. 2

The cavity brickwork was formed in 300mm lifts incorporating galvanised wire ties and shear links where required. The cavity was then grouted and hand tamping was used to consolidate the grout. The top surface of the grout in each lift was kept 35mm below the top of the brickwork so that there was no continuous construction joint through the wall at any point. Once the bricklayer had a little experience of this process work proceeded rapidly with no interuptions. Vertical bars in the walls were lapped to the starters and hoop steel fixed as the brickwork was raised. The backface of the brickwork was painted with bitumen.

This is normal practice for masonry earth retaining walls and prevents ingress of groundwater to the brickwork. It is questionable if this is strictly necessary for underground water tanks.

Waterproofing

The tank, as build, included a waterproof render of SBA (Styrene Butadine additive) mortar. Before this was applied the tank was filled and then emptied to allow any cracking of the brickwork to occur. This was to reduce the strains in the render once the tank was rendered and refilled. The SBA mortar has particularly good waterproof and strength characteristics and is thus ideal for this sort of situation. However, the cost of rendering is obviously a consideration and it may be possible to incorporate the SBA in the grout that forms the core of the wall. Another aspect of waterproofing is the degree of compaction of the grout in the cavity. The wall, when constructed, was brought up in successive lifts of approximately 300mm. The grout, when poured in, had to be hand compacted which is not entirely satisfactory. There are now a number of fluidizing agents that are added to concrete for other reasons (superplasticisers based on sulphonated melamine formaldehyde or naphthalene formaldehyde) that could possible be used to ensure full filling of the cavity. However, their performance with SBA is not known.

Economics

Analysis of costs show that a circular reinforced brickwork tank is cheaper than an equivalent reinforced concrete tank. The major saving is in the cost of formwork. No costings are at present available for rectangular tanks although recently one has been built. However in this case size and site access restrictions made brickwork the only economic choice.

The ratio of costs for circular buried water tanks is estimated to be of the order 1:1·1 (brick:concrete). The comparison becomes even more favourable for above ground tanks where a fair faced concrete finish is often required and there are additional shuttering support costs.

SUMMARY

In order to design and build the tank as described required a mixture of reinforced concrete theory and reinforced brickwork theory. This is an unsatisfactory state of affairs, but is understandable given the present state of the art for reinforced brick structures of this type. The exercise illustrates some areas that need further consideration if this type of structure is to be more widely used. Firstly, the behaviour of brickwork needs to be examined for the kinds of stress systems developed in tanks and then incorporated in design tables so that any designer has the option of using reinforced brickwork. Secondly, the use of additives in grouted cavity work needs to be investigated - possibly the use of waterproofing and superplasticising agents together. Thirdly, the shear behaviour of grouted cavity walls with reinforcement needs a full scale study.

Due to stability problems effecting the excavation sides backfilling the tank started approximately 14 days after bricklaying began. Originally it was intended to fill the tank to test for watertightness before backfilling. So that some monitoring of leakage could take place, four vertical 225mm diameter plastic pipes were placed equally around the tank near the wall face. The pipes were drilled with holes along their length and the top of the pipes projected above ground level. Any serious leakage would then manifest itself as a rising ground water height in the pipes - this could be monitored by a simple arrangement of string with a weight lowered into the pipes.

Fig.2

Fig.3

Fig.4

On completion of the brickwork the tank was left empty for approximately 1 month. After this time the tank was filled to test for structural stability and watertightness. As had been suspected from the start of the project and from previous experience of an earlier brickwork tank[9], the unrendered brickwork was not totally watertight. The tank was emptied, the SBA mortar render applied in two 8mm coats and, after curing, the tank was refilled. No leaks were detected and the water level in the tank remained constant.

Summary of construction sequence

The construction sequence was as follows:

(1) Excavate to firm foundation. Install pipework under base. Lay sand/cement over site blinding to a minimum depth of 75mm.
(2) Fix base steel and wall starters. Set up base shuttering including the formwork for the concrete upstand which forms part of the wall/base junction.
(3) Pour base concrete in one operation
(4) Construct reinforced grouted cavity brickwork in 300mm lifts incorporating vertical and horizontal steelwork as work proceeds.
(5) Paint earth face of the brickwork with bitumen
(6) Backfill to ground level in equal layers not greater than 1m deep. Incorporate inspection pipes in backfill
(7) Construct 215mm thick plain walling above ground level.
(8) Fill and empty tank. Apply SBA render in two coats. Total thickness of render not less than 16mm.
(9) Fill and monitor water level

Progress of construction can be seen in figs. 2,3 and 4.

ACKNOWLEDGEMENTS

This work is published with the permission of the Directors of Structural Clay Products Ltd.

REFERENCES

(1) British Standard Code of Practice, BS.5337 : 1976. "The structural use of concrete for retaining aqueous liquids". B.S.I. London
(2) British Standard, BS.4449 : 1978. "Hot rolled steel bars for the reinforcement of concrete". B.S.I.London
(3) British Standard Code of Practice, CP.110 : Part 1 : 1972. "The structural use of concrete". B.S.I. London
(4) "Reinforced concrete reservoirs and tanks", G.P. Manning, Cement and Concrete Association 1972.

(5) British Standard Code of Practice CP.111 : Part 2 : 1970. "Structural recommendations for loadbearing walls". B.S.I. London

(6) British Standard Code of Practice CP.114. "The structural use of reinforced concrete in buildings, 1969." B.S.I. London

(7) "Reinforced Brickwork : vertical cantilevers 2" Structural Clay Products Limited, Hertford, U.K. 1976.

(8) British Standard 3921, "Clay Bricks and Blocks", B.S.I. London.

(9) "Design of a prestressed brickwork water tank", Structural Clay Products Ltd., Hertford U.K., 1975

7. Reinforced brickwork in the Hyde Park Methodist Church, Leeds

R. E. BRADSHAW, MSc, MICE, FIStructE, MConsE, Director, Bradshaw Buckton & Tonge, Leeds

SYNOPSIS. The reasons for using reinforced brickwork in a new Methodist church are given together with details of the elements reinforced. The technique was found to be well within the capability of a builder used to traditional materials and assisted in achieving a low cost, and relatively maintenance free building.

INTRODUCTION

1. The Hyde Park Methodist Mission serves the needs of four churches that merged their congregations to form one new church in 1975 and caters for religious, social and educational activities. Accommodation includes a church, a large multi-purpose hall, coffee bar and dining room, a variety of smaller rooms and a caretaker's flat. (Fig. 1, 2, 3).

2. Brickwork was chosen by the architect for the major part of the scheme for a variety of reasons as noted below and the adoption of reinforced brickwork for certain elements was a natural development in achieving the architectural requirements economically.

(a) A maintenance free building was required and an impervious smooth faced brick was selected for both internal and external walls.

(b) Mass was required to ensure adequate sound reduction for noise from outside the building and also to minimise noise transfer from one area of the building to another.

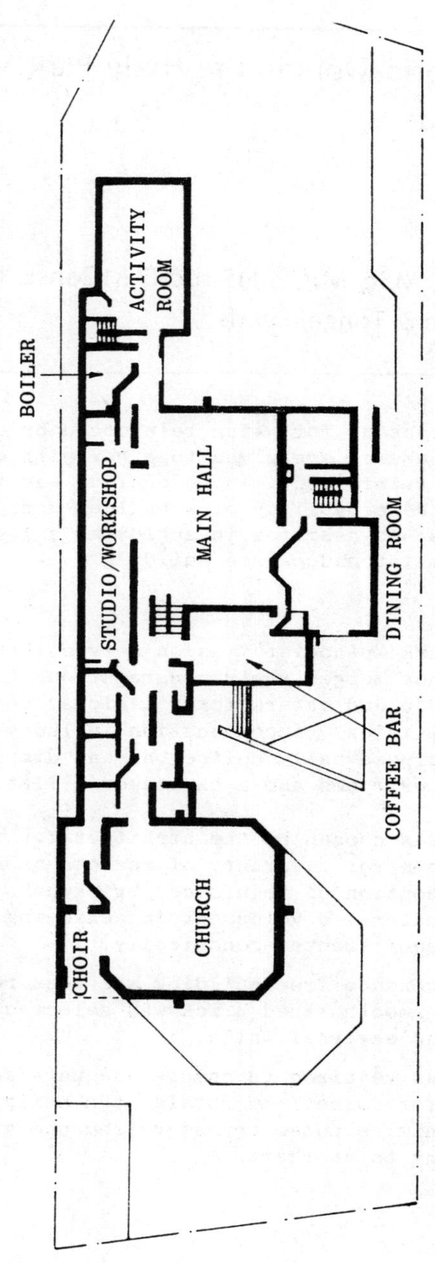

Fig. 1. Ground Floor Plan

Fig. 2. View to Main Entrance. Church is on the left; Hall in centre and chimney to the right

Fig. 3. Rear view

(c) Brickwork has a unifying quality when used in association with other materials such as timber and quarry tiles and this assisted in the design requirement to provide variety and interest in the completed building.

(d) Brickwork represents a sense of permanence which, in a world of rapid change, was considered important especially as symbolically the church is founded on a rock.

(e) Impervious brickwork was considered to be resistant to potential vandalism and would be easier to clean than more porous materials.

(f) The building was considered to be a low cost scheme and this, together with the above requirements was achieved through the brickwork fulfilling a number of roles.

SITE INVESTIGATION AND FOUNDATION DESIGN

3. Terraced housing on the site was demolished and boreholes revealed fill of varying depths with a maximum depth of 4m beneath the church. Alternative foundation schemes were prepared and costed, including piling.

4. The cheapest scheme was adopted which comprised reinforced concrete pad bases at a depth of 4m beneath the perimeter walls to the church. Mass brickwork piers, 450 square and 450 x 462, were then built off these pads to the underside of reinforced concrete perimeter ground beams. These were 375 wide x 550 deep and were designed to act compositely with the brickwork walls above using a maximum bending moment of $\frac{WL}{25}$ (Ref.1). All the shear forces were carried by the ground beams. Loads were approximately $9^{T}/m$ and main bottom reinforcement was 3no. 25 dia. bars.

5. The church hall and other accommodation was founded on an RC raft slab on compacted hardcore.

6. Sulphate resisting cement was used for all substructure concrete and mortar due to the presence of sulphates in the fill.

7. A 30m length of walling to an activity room beneath the caretaker's flat served as a retaining wall due to the natural slope on the site.

At the design stage (1972), it was recognised that a reinforced brickwork retaining wall would be cheaper than RC although little published information was available on comparative costs of the three main types of grouted cavity, pocket and Quetta. A 337 Quetta bond wall with 20 dia. bars at an equivalent average spacing of 337 was used. Pockets were at 225 c/cs and at every third pocket the bar was omitted (Fig.4).

THE CHURCH

8. The church was initially planned as 15m square with splayed corners 3m long and 9m high. The walls were 375 cavity construction, fair faced brickwork both internal and external with a 225 inner leaf and 100 outer leaf.

9. Alternative ways of achieving stability of the brickwork panels and the church as a whole were considered, including steel framing and several brick pier schemes.

10. The size of the church was later reduced to 13m square and for the final scheme a central reinforced brickwork pier or fin 337 x 450 projection, was introduced at the centre of each of the large brickwork panels. (Fig.7).

11. The piers were designed to span 9m vertically from ground beam to roof. The roof comprised plywood/timber beams spanning 13m and was stiffened to provide diaphragm action and lateral support to the tops of the reinforced brickwork piers by a 1.3m wide horizontal band of plywood adjacent to the perimeter walls.

12. Initially, a number of small openings in certain of the main walls were required in order to accommodate glass blocks. These were to enhance the natural lighting within the church and add variety to the brickwork walls.

13. The final design incorporated a larger number of windows, larger in size than originally envisaged. (Fig.8). A re-appraisal of the $\frac{WL}{25}$ composite design approach was made and reinforcement was incorporated over the openings.

CHURCH HALL

14. The church hall is approximately 14m by 10m with a mono pitch roof and is abutted on three sides by ancillary accommodation. The floors and roofs were designed to provide lateral support and stability to the wall panels and the complete hall.

15. On the 4th side, a reinforced brickwork pier similar to those for the church was provided.

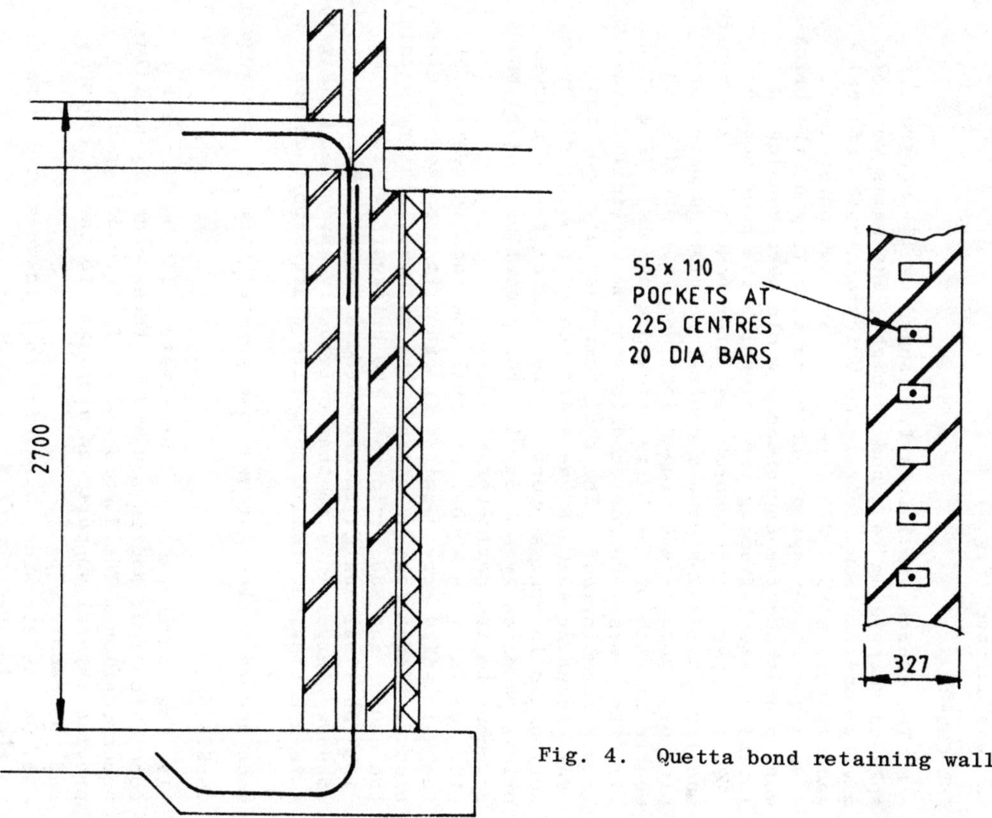

Fig. 4. Quetta bond retaining wall

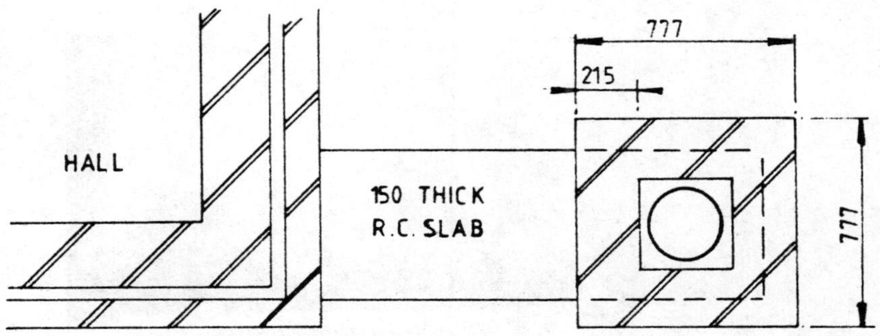

Fig. 5. Plan on chimney

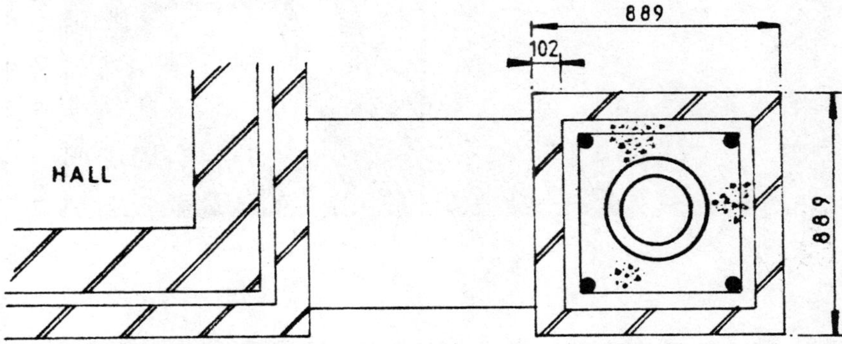

Fig. 6. Alternative plan for chimney

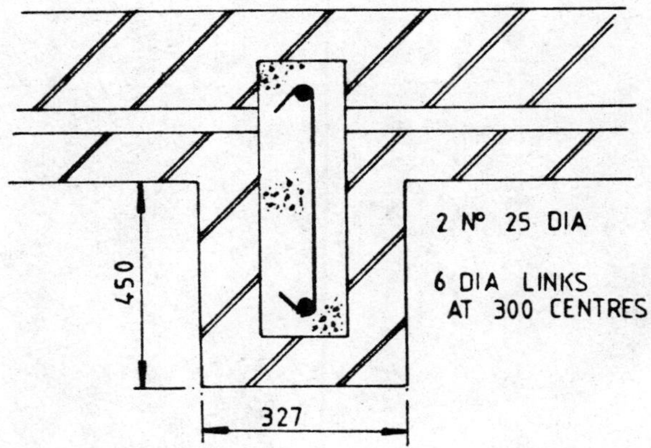

Fig. 7. Typical pier to church and hall

Fig.8. Wall to Church showing large windows and pier

Fig. 9. Brickwork chimney. Note R.C. tie 3.3m in from top

CHIMNEY

16. A brickwork chimney of total height 12m was required to house a stainless steel liner approx. 250 dia. The chimney was constructed in 225 brickwork and was 800 square on plan although an earlier scheme incorporating 102 brickwork and vertical reinforcement was also considered. (Figs. 5 and 6).

17. Lateral support to the chimney at a height of 8m was provided by a 150 thick RC slab cantilevering from the corner of the hall where the brickwork had been suitably reinforced. (Fig. 9).

18. The small void between the stainless steel liner and the brickwork was filled with concrete and the top 3.3m of the chimney was adequate as a cantilever under wind loading without developing tension in the brickwork.

19. The lower part of the chimney was integral with intersecting brick walls for stability and certain of the brickwork courses were reinforced with brickforce to accommodate any differential movement that might arise due to any thermal gradient local to the boiler.

OBSERVATIONS

20. The work was carried out by a local small/medium sized traditional builder who coped well with the brickwork construction. The use of reinforced brickwork appears to be well within the capability of a builder experienced in traditional materials.

21. The adoption of reinforced brickwork for critical structural areas of a project can eliminate the need for structural steel or reinforced concrete and this can result in a simplified construction process and often reduced building costs.

ACKNOWLEDGEMENTS

Architect: Brooks Thorp & Partners
Structural Engineer: Bradshaw Buckton & Tonge
Structural Engineer for Timber Roofs : Ian H Paxton & Associates
Quantity Surveyor: A Leslie Heaton
Contractor: J W Dufton & Sons Limited

REFERENCE

R H Wood and L G Sims. A tentative design method for the composite action of heavily loaded brick panel walls supported on reinforced concrete beams. Building Research Current Paper 26/69.

8. Post-tensioned, free cantilever diaphragm wall project

W. G. CURTIN, G. SHAW, J. K. BECK and L. S. POPE,
W. G. Curtin and Partners, Liverpool

SYNOPSIS. The paper describes the first known application of the uncompleted research on highly stressed post-tensioned diaphragm walls (ref.1) which followed an earlier development of the authors' practice (ref.2).

INTRODUCTION

1. The salvation Army needed a new hall for religious services, the hall was to be 25 metres long x 15 metres wide x 8.5 metres high as shown in fig.1. Since the cost budget was extremely tight it was essential to produce a highly economical solution. It was known, from much experience on similar tall single-storey structures that a diaphragm wall would provide this - but there was a snag. Though the dimensions of the hall were not exceptionally large (the authors have successfully designed much larger halls using the diaphragm wall technique) the top of the wall could not be propped, thus forming a 'propped-cantilever' as common on other projects, since the client required a clerestorey window, shown in fig.2, running round the top of the wall. A propped cantilever wall has a base bending of $\frac{ph^2}{8}$ (where p = wind pressure and h = height of wall) but a free cantilever has a base moment of $\frac{ph^2}{2}$ - i.e. 4 times that of a propped cantilever. The stability moment (or resistance moment) at the base is the product of the wall's own weight times its lever arm (see ref.3). The solution of increasing the lever arm (thus increasing the overall depth of the wall) was rejected both on the grounds of cost and its unacceptability to the architect. The economic solution, but for the recent research, would have to be the incorporation of a steel frame and attendant cladding, insulation and lining. Fortunately the early

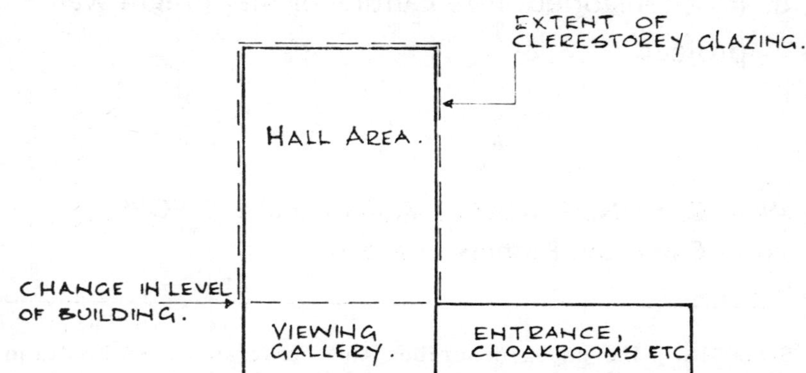

FIG 1
PLAN ON HALL.

Fig. 2 VIEW SHOWING CLERESTOREY WINDOW

results of the research (ref. 1) had just been produced at the time of the preliminary design stage of the project. The alternative to increasing the lever arm is to increase the 'own weight' by heavy prestressing, this had been success -fully proven by the research and it was decided to design a post-tensioned diaphragm wall.

DESIGN CONCEPT

2. The moment of resistance, ignoring any tensile resistance, at the d.p.c. of a diaphragm wall is W x a (where W = weight of the wall and a = lever arm). The moment of resistance within the height of the wall is f x Z where f = stress and Z = section modulus). W can be increased slightly by increasing the overall depth the wall - so can 'a' and Z, but this is usually uneconomic. But both W and f can be increased by post-tensioning to cope with the quadrupling of the bending moment. The wall does not then need thickened leaves, or increased overall depth, and structural masonry can again prove more economic that structural steelwork.

3. The normal governing bending value of 'f' in plain brickwork is its pathetically low tensile strength but the application of prestressing can easily increase the bending resistance tenfold (ref.4). With such an increase in resistance it is simple to cope with the quadruple increase in the bending moment and resultant bending tensile stress. By keeping the diaphragm wall to the normal overall depth it has, easily, a sufficiently high radius of gyration (which, of course reduces the slenderness ratio) so that it can accommodate the higher axial compressive stress induced by prestressing.

THE STRUCTURAL DESIGN

4. The diaphragm wall section as shown in fig 3 was chosen after consideration was given to both the architect's requirements, at low level in respect of window arrangement (see fig 2) and the particular bending moments induced by the wind forces at that level. The introduction of the post-tensioned force created a condition whereby no tensile stresses existed at the base of the wall under full wind load. The bending tensile stresses due to wind load, without prestressing, were far in excess of the wall's capabilities even if full tensile stress throughout the wall were permissible - which is not normally the case at d.p.c. level. The post-tension force was achieved by the introduction of 2 No. 32 mm diameter Macalloy bars positioned within the section of the diaphragm void in fig.3 and anchored as shown in fig.4.

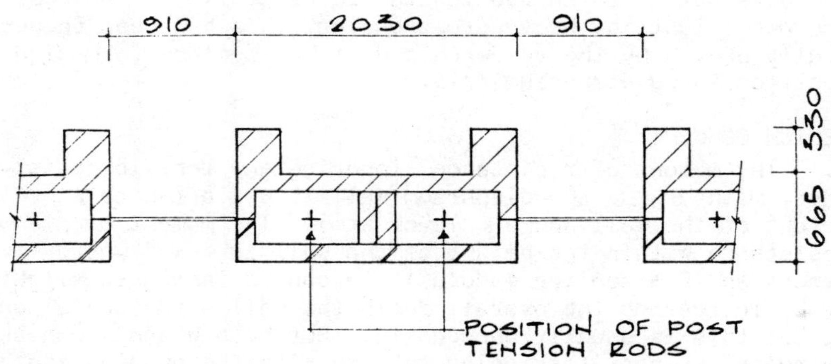

SECTION AT LOWER LEVEL

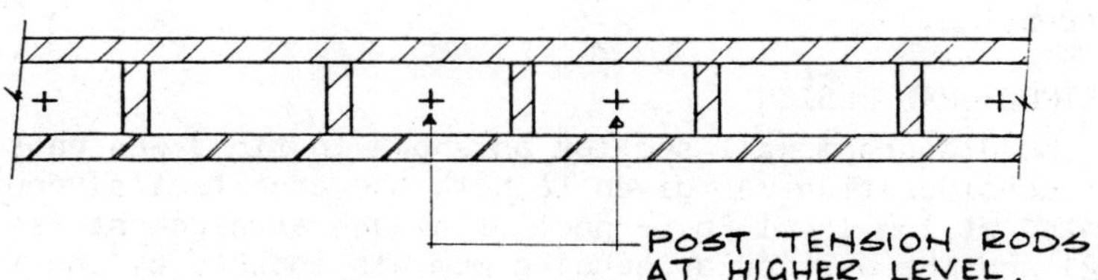

SECTION AT HIGHER LEVEL.

FIG 3

DIAPHRAGM WALL - SECTIONAL PLANS

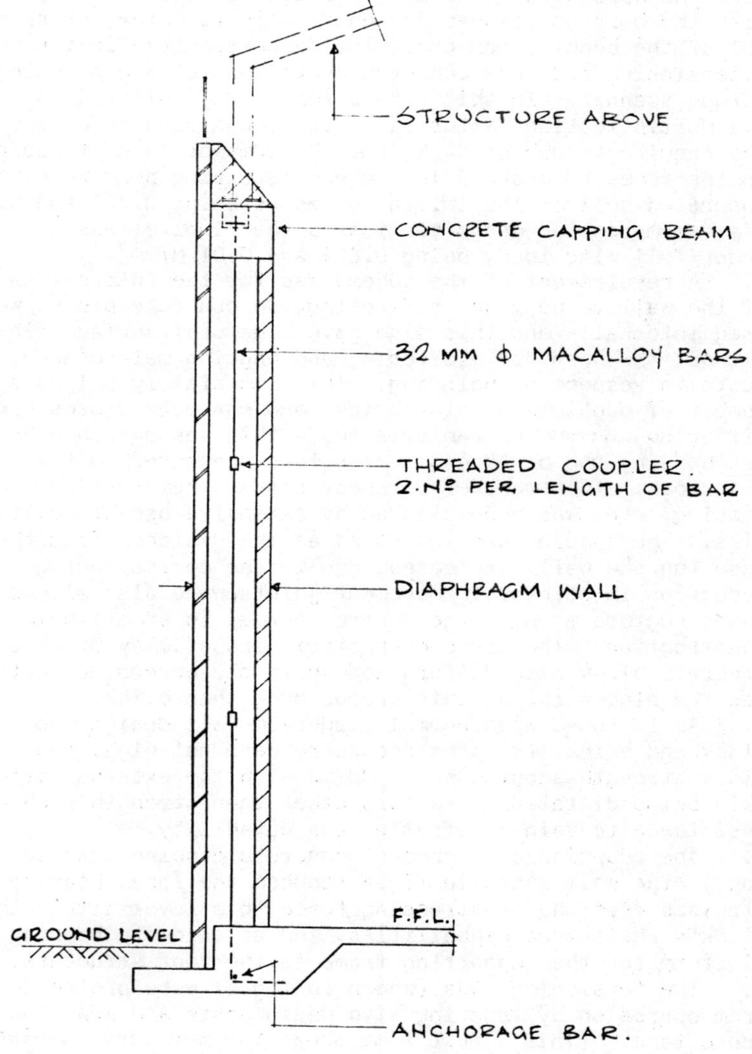

FIG 4.

OVERALL SECTION - ANCHORAGE DETAILS.

5. The bars were tensioned to induce a force of 100 kN with the bars positioned concentrically to cater for reversal of the bending moments. The force required in the post - tensioning rods was achieved by the use of a conventional torque spanner. In this case a Torque Multiplier with a 5:1 Multiplication factor (see fig.5) was used to achieve the require torque of 64 kg fm. On completion of torquing up the recessed pockets in the concrete ring beam were concreted solid. The initial stresses being 0.713 and 0.05 N/mm^2 compression and after losses the final stresses, under full wind load, being 0.673 and 0.01 N/mm^2.

6. A requirement of the scheme was for the internal face of the wall to be light reflecting, so concrete blocks were used internally and this also gave a natural surface finish without the need for plastering and ongoing maintenance costs in respect of painting. This immediately raised a number of problems as clay bricks and concrete blocks have differing movement characteristics - this was overcome by <u>not</u> bonding the block cross-ribs into the outer leaf of brickwork. The transfer of shear stress, compression loading, etc. was accomplished by extensive use of cavity ties. The spacing was increased as the distance from the base (up the wall) increased, due to the corresponding reduction in stresses. Movement joints were also placed in the structure at half the centres normal in an all-brick construction. The creep characteristics of clay brick and concrete block also differs and again the governing factor was the blockwork, as this creeps more than brick.

7. As is found with normal diaphragm wall design the block and brick strengths necessary were not high, the block strength adopted was 7 MN/m^2 with the external brick skin being dictated by factors other than strength such as resistance to rain penetration and durability.

8. The adoption of a precast concrete capping beam to the top of the wall shown in fig.4 reduced the local bearing stresses from the prestressing force to a level within the block's resistance capabilities, and an adequate bearing platform for the supporting frame to the roof structure.

9. The tensioning rods (shown in fig.4) were protected from corrosion by treating with Denso paste and wrapping in Denso tape. (This figure also shows the way service pipes such as rainwater pipes can be accommodated within the diaphragm section. Threaded coulpers, (shown in fig. 4) were adopted to extend the length of the tensioning bars.

Fig 5 TORQUE MULTIPLIER

Fig 6. BAR PROTECTION AND COUPLER

CONSTRUCTION

10. The contract for the hall and ancillary buildings was £200,000. Since this was a relatively small project it was well within the capability of a small local builder. But the building of a tall post-tensioned diaphragm wall with its image (false) of complex high technology might not appear to be within the capacity of such small building firms - but this is not so in practice. As far as the builder is concerned he is building a wide cavity wall with ribs not cavity ties and rods in the cavity void which are "screwed up" later. Certainly there is a need to provide the contractor with clear and simple working drawings and a good specification. Guidance is, on the whole, lacking in both these requirements. Experienced engineers should have little problem in providing the site with the adequate working drawings. There is a problem however with the specification. Most specifications only deal with normal brickwork for simple walls of normal room heights and there is not, to the authors' knowledge, a comprehensive specification for modern structural brickwork. Fairly certainly there is none for tall post-tensioned diaphragm walls of brick and block construction. The authors practice has, over the years, acquired the knowledge and experience to provide guidance on both specifications and working drawings and will publish this in the near future (ref.5 and 6). There is little doubt that other engineers working in this new field of construction technology will also publish guidance in the not too distant future.

11. No problems were encountered by the builder during the construction, the contract was completed on schedule, within the budget price, with a good standard of workmanship and the client is pleased with the result.

CONCLUSIONS

12. The first, early (almost premature!) application of post-tensioned diaphragm wall research augers well for the future development of the fuller potential of the technique. It may also perhaps further encourage cooperation between research and practice which has proved so fruitful in this example.

REFERENCES

1. CURTIN AND PHIPPS 'Behaviour of Post-Tensioned Diaphragm Walls' IBMAC 1982.

2. SHAW G. 'Post-Tensioned Brickwork Diaphragm subject to severe Mining Subsidence' ICE/ISE Symposium 1982.

3. CURTIN, SHAW, BECK AND BRAY 'Design of Diaphragm Walls for Tall Single-Storey Structures' BDA.

4. CURTIN, SHAW, BECK AND BRAY 'Post-Tensioned Brickwork' IBMAC 1982.

5. CURTIN, SHAW, BECK AND PARKINSON 'Structural Masonry Detailers Manual' Granada Publishing - Technical Division, London , 1983.

6. CURTIN, SHAW, BECK AND PARKINSON 'Specificiation and Construction of Structural Masonry' Granada Publishing - Technical Division, London. (in preparation).

9. Reinforced and prestressed masonry in agriculture

J. P. DRINKWATER, BTech, Engineer and R. E. BRADSHAW, MSc, MICE, FIStructE, MConsE, Director; Bradshaw Buckton & Tonge, Leeds

SYNOPSIS. Reasons for the use of reinforced masonry in agriculture are given together with a cost comparison of the main types of retaining wall. Several case studies are included.

INTRODUCTION

1. This paper is concerned with stimulating interest in reinforced and prestressed masonry. It is not intended to give calculated examples, nor to comment on design principles but more to create confidence in those who have not used this adaptable construction medium.

2. The examples given are all concerned with reinforced or prestressed brickwork, not because of any prejudice against blockwork, but simply because brickwork has been demanded from planning or other considerations.

3. Brickwork can obviously be reinforced to form any type of load bearing member but perhaps the easiest and most efficient use is in retaining walls, and it is in this mode that the demand has come from the agricultural world.

4. The word retaining is used in the widest sense and the retained material could be anything from natural soil to silage, slurry, water or food crops, etc. Designs have been carried out for all the above conditions and in every case reinforced masonry has been economic and also fulfilled the aesthetic and durability requirements.

5. Whilst economy of construction is directly linked to unit cost, more importantly it is also dependent on "buildability". It is very easy to design a wall with the minimum of materials whilst making it practically impossible either to build or achieve the necessary level of workmanship.

6. The general reluctance (or ignorance) on the behalf of most engineers to design in reinforced masonry is also

reflected in many contractors wariness of pricing and building. However, engineers who have successfully completed projects using reinforced masonry have realised that it is neither difficult to design or understand, and once contractors have understood the simple techniques employed in construction, their tender figures are more realistic second time around.

REASONS FOR MASONRY IN AGRICULTURE

7. In order to keep costs down farmers very often like to use their own labour for construction work and anyone who can lift a trowel is press ganged into building some masonry. The subject of buildability becomes even more important in these cases as simple details mean that mistakes are less likely to occur on site and site inspection visits can be made less frequently.

8. The retaining walls are often required to enclose an existing open sided building to form a covered store and masonry has proved adaptable in building around obstructions and within spaces with limited access, etc. The types of retaining wall generally used are mass, grouted cavity, quetta, post tensioned diaphragm, pocket type, faced and unfaced concrete.

9. These are all used in various applications although the mass type is not recommended in anything but very minor walls due to the high risk of failure due to accidental damage. The quetta type is not often used as it is complicated to construct with few advantages over a grouted cavity wall.

COMPARATIVE COSTINGS

10. In order to make the most efficient use of the available types of wall, the designer must have a knowledge of the overall costs involved and also the different areas in which the money is being spent.

11. To this end, Messrs. George Armitage & Sons have financed a costing exercise using the above wall types and the results confirm that reinforced brickwork can be economic if used in the appropriate conditions.

12. A series of designs were carried out with the wall height varying from 1m up to 5m in 1m increments. The types considered were, reinforced concrete, brick faced R.C., pocket, grouted cavity and a mass wall. The study of quetta bond, post tensioned and reinforced diaphragm and blockwork walls are scheduled for a future date.

13. The costings were carried out by a firm of Quantity

Surveyors using billed rates and as such may not reflect the buildability aspect of the brick walls. Experience has shown that many builders prefer to work with bricks rather than with concrete and price accordingly.

14. The comparative costs are shown in Fig.1. It should be noted that each costing curve should be represented as a band say ± 10-15% about the line to take account of:-

(a) Whether a contractor is geared up for one form of construction compared to another.

(b) Size of project.

(c) Degree of repetition, etc.

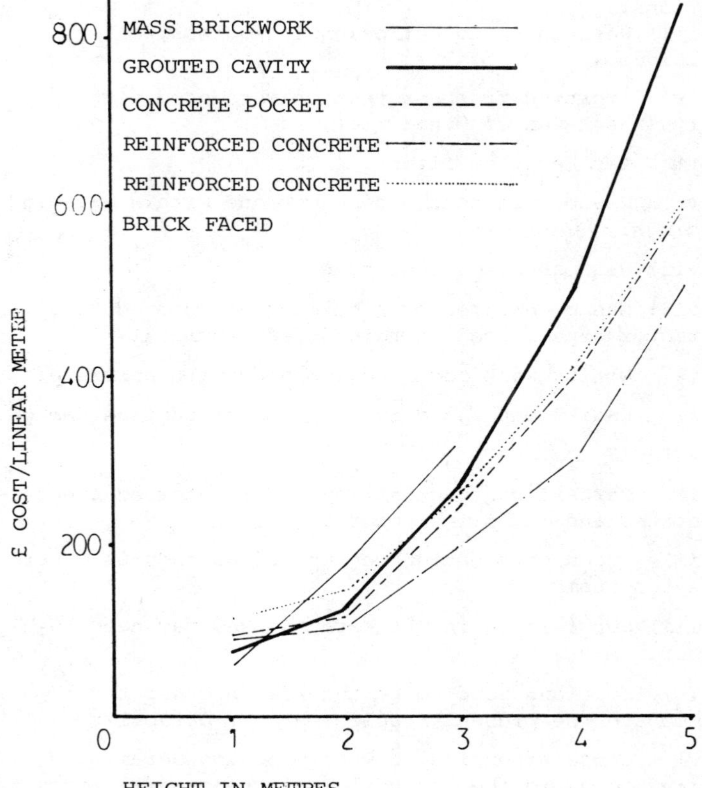

Fig. 1. Comparative costs of various types of retaining wall based on billed rates for a medium/large project and assuming re-use of shuttering.

15. It can be seen that reinforced masonry is competitive with brick faced reinforced concrete walls but that the difference reduces as the height of the wall increases.

16. At a wall height of 4 to 5m factors such as project size, repetition and contractor preference will decide which type of wall is the cheapest, together with a detailed consideration of the advantages and disadvantages given below.

ADVANTAGES

(i) Reduced maintenance.

(ii) Ease of sterilising surfaces with acidic solutions.

(iii) Resistance to corrosion in aggressive conditions.

(iv) Improved frost protection when used with a cavity (insulated or otherwise).

(v) Aesthetic qualities.

(vi) Useable in confined spaces and within existing buildings.

(vii) High abrasion resistance.

(viii) Can be erected by a builder as opposed to a contractor experienced in reinforced concrete.

(ix) No formwork required compared with concrete.

(x) Easily made good in the case of surface damage.

DISADVANTAGES

(i) Certain contractors may be "reinforced concrete" orientated and may price accordingly.

(ii) Speed of erection may not be as fast as other package systems.

(iii) Supervision in the early stages may need to be greater.

(iv) Sections tend to be thicker in order to accommodate the present requirements of durability.

(v) Close attention to waterproofing details required to avoid the unsightly effects of efflorescence and water staining.

(vi) The requirement for different sands in mortar mixes and concrete mixes can cause control problems.

CASE STUDIES

17. Three projects follow which incorporate reinforced or prestressed masonry in agriculture and a brief description of each is given.

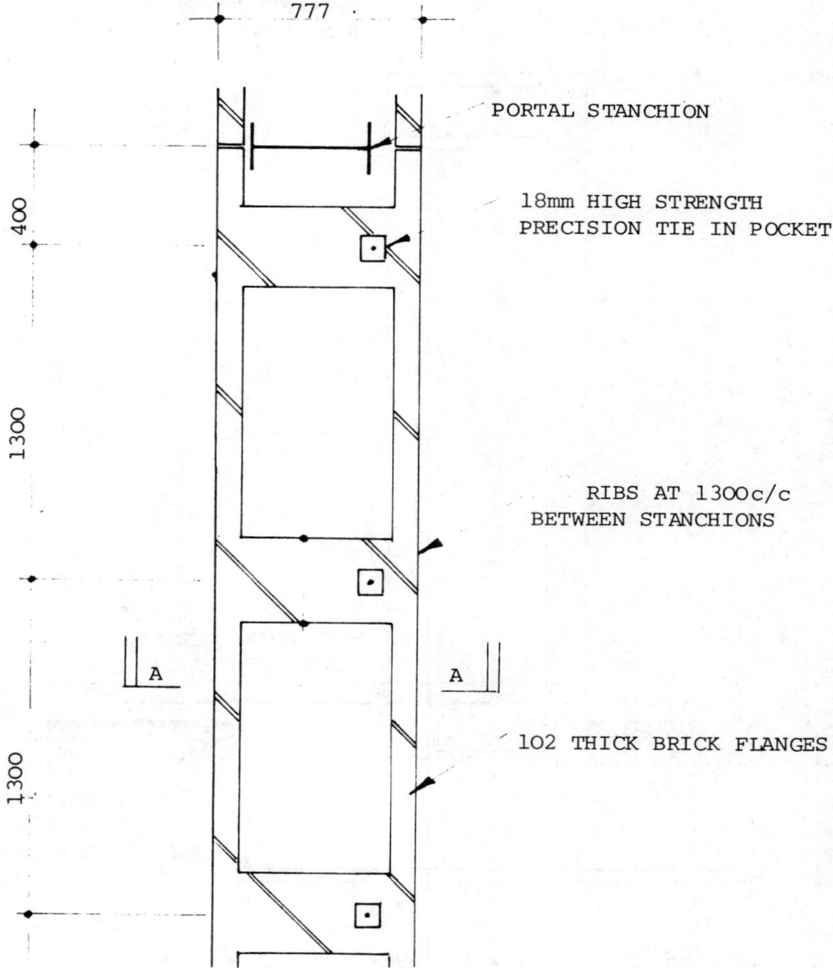

Fig. 2. Multi-purpose farm building. 30m square steel portal frame. 2.4m high post-tensioned brickwork retaining wall. Used for storage of a variety of crops and as a workshop.

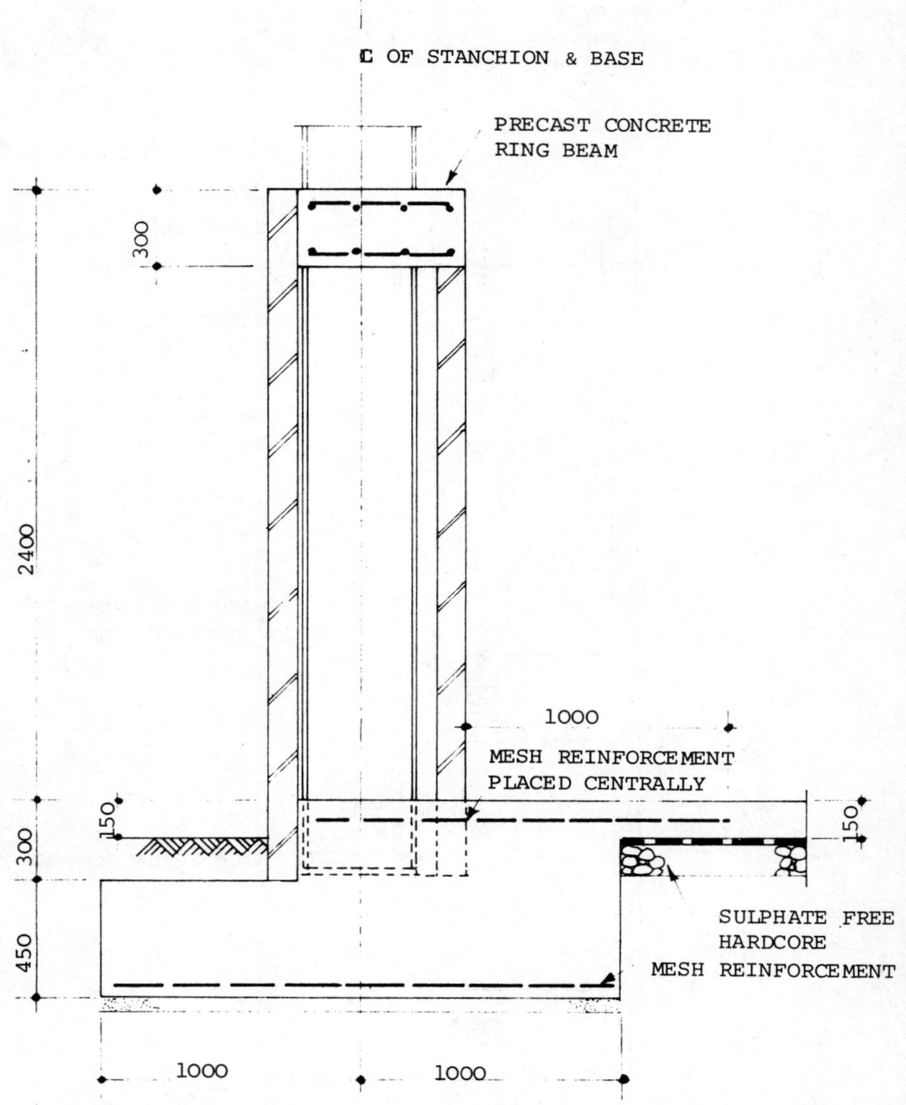

Fig. 3. Typical Section A-A

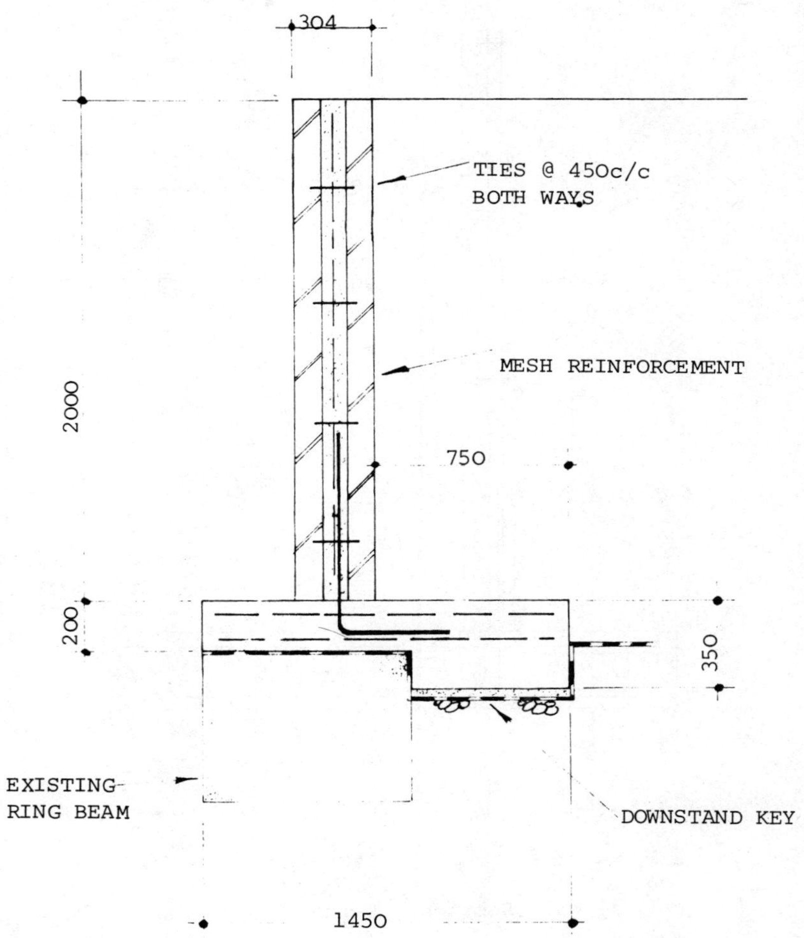

Fig. 4. Typical Wall Section. Conversion of an existing Timber Framed Building to a grain store. Farmer had little previous experience in building works. Scheme designed to retain existing foundations and timber posts and phased to ensure minimum disruption.

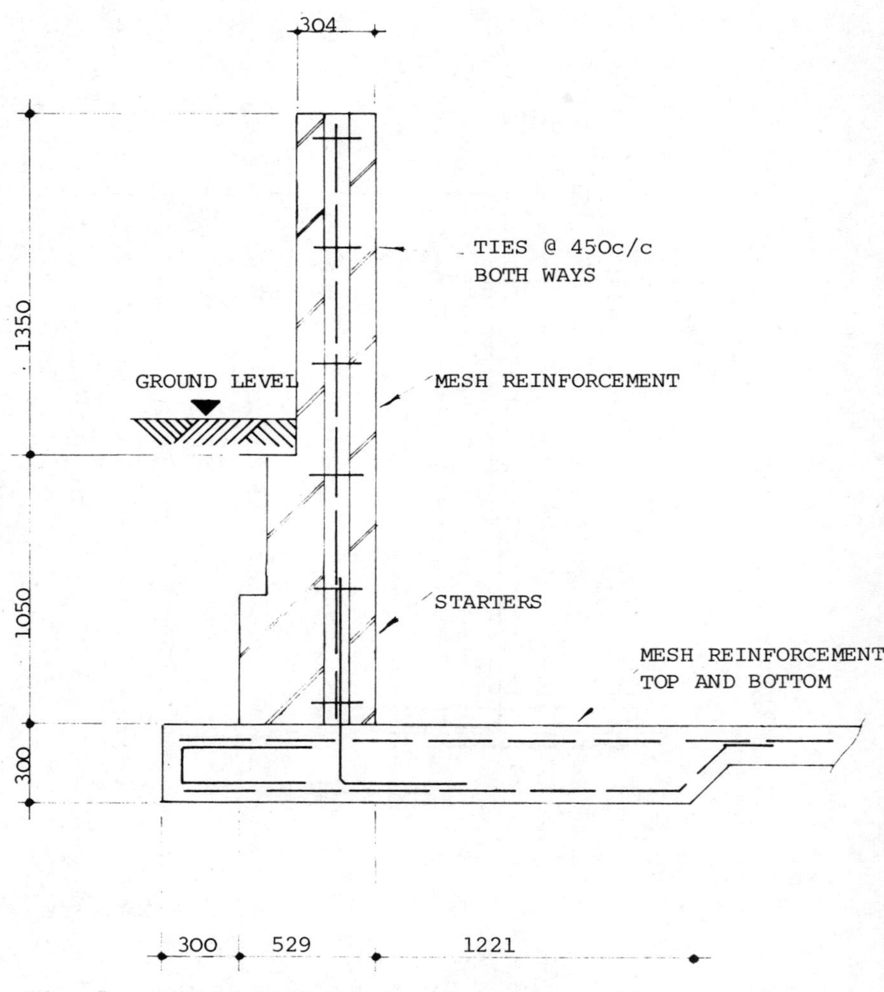

Fig. 5. Typical Wall Section. Large slurry tank. 30m x 14m x 2.4m deep.

Acknowledgements

The comparative costing exercise was financed by George Armitage and Sons plc and prepared by Rex Proctor & Ptnrs.

The case studies relate to projects carried out for Weasenham Farms Ltd and also Mr Caldwell.

10. Prestressed blockwork silos

T. J. S. MALLAGH, MA, MAI, FIEI, MConsEI, Consultant,
Mallagh Luce & Partners, Dublin

SYNOPSIS. An early method of construction of tall silos of prestressed concrete blockwork is described. The bins, circular in form, were built of 4½" nominal blockwork and the circumferential wires stressed by pulling them together in a vertical direction alternatively upwards and downwards by a simple lever tool. These wires at their displaced points were then tied together by soft-wire ties.

INTRODUCTION.

1. About the year 1950 many reinforced concrete grain storage silos were being repaired by the Gunite (sprayed mortar rendering) process and there was evidence that Gunite would resist weather attack better than average quality poured concrete. These considerations led to the idea that a method of construction which involved an external layer of Gunite initially would have much to commend it.

2. Concrete blockwork with Gunite external rendering was a fairly obvious solution. Circular bins were envisaged as only ring tension would have to be resisted; with square or hexagonal bins, reinforcement to resist both tension and bending is necessary. Owing to the relatively high shrinkage characteristic of concrete blockwork, it was felt that circumferential prestressing of the walls should be adopted, instead of normal reinforcement.

METHOD OF CONSTRUCTION.

1. At the time it was considered that conventional prestressing, involving jacks and anchorages, would have certain disadvantages for a circular structure, particularly the inherent disadvantage of having to allow for large friction losses on pulling a wire round a large arc. (In

fact later work proved that such misgivings were unfounded.) These considerations led to the idea of tensioning the wires by the application of a relatively small transverse force at several closely spaced intervals round the circumference of a circular bin.

2. The concept was developed to give the method illustrated in the diagram, figure 1.

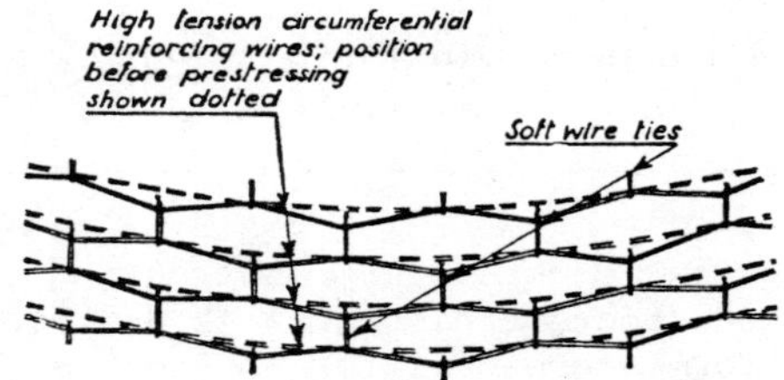

Fig. 1. Arrangement of reinforcing

The lever used to draw the wires together is illustrated in figure 2.

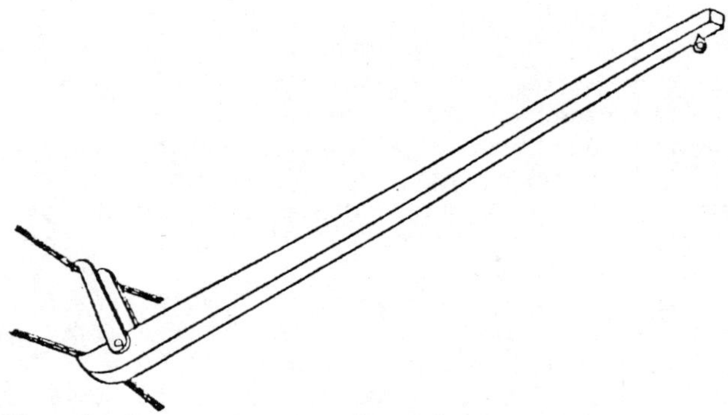

Fig. 2. Lever for drawing wires together

As can be seen, the main circumferential wires were pulled together at staggered centres and held in their deflected positions by soft wire ties. The ties were of the double hairpin form and were closed in position by several twists. The twisting was done using a simple tool comprising a length of ½" diameter steel rod with a hook at one end and a handle at the other. To determine the stress induced in the circumferential wires, the pull on the end of the stressing lever was measured.

FIRST SILO.

1. The first silo of significant size using this method of construction was one of 2,000 ton capacity built for the Dublin North City Milling Company in 1952. The walls were 4½" nominal blockwork and the circumferential reinforcing comprised two 10 SWG wires at 9" centres.

2. The maximum design tension in these wires was 112,000 lb per sq inch which would induce a compressive stress in the blockwork of 78 lb per sq inch. With the method of prestressing adopted, the stress in the circumferential wires varied from place to place. The minimum measured before the application of the Gunite rendering was 66,200 lb per sq inch. It was reckoned that creep and shrinkage would reduce this to 60,000 lb per sq inch. This would induce a compressive stress in the blockwork of 42 lb per sq inch against a calculated tensile ring stress of 38 lb per sq inch due to the pressure of the stored grain. Tensile stress in the blockwork was therefore prevented.

3. With tall silos, one has also to consider the compressive load in the blockwork due to a proportion of the weight of the stored grain being transmitted to the walls by friction, plus the wind induced load on the lee-ward side. In most cases 4½" nominal blockwork will adequately carry such loading.

SUBSEQUENT SILOS.

1. Following the building of the silo for the Dublin North City Milling Company others were constructed using the system as described above, notably two of 3,000 ton capacity with 20 foot diameter bins, one at Bristol and the other at Hull.

2. It was, however, realised that the system of pre-stressing as described above had certain disadvantages. It was time consuming as the stress had to be built up gradually working round the circumference and from tendon to tendon. Stress was apt to be lost due to the wire ties untwisting slightly. This, of course, could only occur prior to the shooting of the Gunite. Once the Gunite had set no further relaxation in this manner was possible. However it was essential to have a dedicated engineer or inspector to see that the minimum requisite prestress was present in the wires before applying the Gunite.

Latterly stressing the circumferential reinforcing has been carried out by applying an axial pull using Gifford-Udal jacks and single wire anchorages.

TESTS.

1. Using strain gauges, an investigation of the stresses in the reinforcement and blockwork of a silo, comprising 10 bins 30 feet in diameter by 80 feet high, was undertaken. The circumferential reinforcing comprised .276" galvanised high tensile wires at 3" to 4½" centres with ¼" vertical mild steel bars at 6" centres between the .276" wires and the concrete blockwork.

2. Stressing and anchoring the .276" wire was carried out at one point in the circumference using 2 jacks pulling in opposite directions. The jack pull had therefore to be transmitted through 180° with consequent friction loss due to the movement of the galvanized wire across the ¼" mild steel bars. For design purposes the load in each wire at the jacking points was taken as 9,400 lb on release and a loss of 5,200 lb over the arc of 180° was calculated using a coefficient of friction of 0.25. Design also allowed for a further loss of 1,400 lb due to creep and shrinkage of the concrete blockwork. The minimum design prestress in the wire was therefore 9,400 lb less 6,600 lb, i.e. 2,800 lb.

3. Strain gauge measurements carried out after a few days stressing indicated a minimum load of 6,800 lb, 4,000 lb more than that calculated. Strain gauge measurements also indicated that the load of 6,800 lb was virtually constant over 360° and did not vary from a maximum at the anchorage to a minimum at 180° as one would expect.

4. Experiments were carried out stressing from one end of a wire and measuring the load transmitted through 360°. This indicated a coefficient of friction of 0.19 effective immediately after jacking. After a few days the load in the wire became virtually constant and consequently friction loss in practice would be even less than that based on a coefficient of friction of 0.19.

5. The maximum stress in the concrete blockwork due to prestressing was that due to the load in the circumferential wire on jacking and before release. This was 10,200 lb which produced a compressive stress of 825 lb per sq inch in the blockwork. At the minimum measured load of 6,800 lb in the wire the stress in the blockwork was 550 lb per sq inch. Strain gauge measurements on the blockwork indicated that pressure from the stored contents induced a tensile stress of 200 lb per sq inch leaving a residual compressive stress in the blockwork of 350 lb per sq inch.

6. The blockwork at the bottom of the silo carried also the load due to the weight of the contents transferred to

the walls by friction. If the whole of the weight of the contents was uniformly carried by the blockwork it would be stressed to 960 lb per sq inch. Strain gauge readings indicated that a large proportion of the weight of the contents was in fact transferred to the walls. With stresses of the order of the figures given above a high quality concrete block and mortar are necessary. Furthermore good quality concrete and mortar are necessary to prevent excessive shrinkage. Blocks having a 28 day strength of not less than 3,000 lb per sq inch when tested on edge were specified. The mortar for jointing was 1 : 2½ cement/sand with plasticiser having 28-day compressive strength of 4,000 lb per sq inch (4" test cubes).

References:

1. MALLAGH, T. J. S., and MADDOCK, R. W., An Unusual Grain Storage Project. Transactions of the Institution of Civil Engineers of Ireland, 1953-54 Vol. 80.

2. MALLAGH, T. J. S., Concrete Silos. Transactions of the Institution of Civil Engineers of Ireland, 1957-58 Vol. 84.

11. Post-tensioned brickwork diaphragm subject to severe mining settlement

G. SHAW, FIStructE, MConsE, W. G. Curtin and Partners, London

SYNOPSIS. The paper describes the building and the site conditions including the mining information upon which the structure design was based. Alternative schemes were considered and the adopted solution is believed to provide the first ever use of post-tensioned diaphragm walls, acting compositely with the cellular raft foundations, to withstand the considerable subsidence movements to which it has been subjected. The acheivement of constructing the building within a tight programme to be in advance of the effects of the subsidence are discussed and the monitoring of the movements of the building and subsequent inspections of negligible subsidence damage are described.

THE BUILDING

1. Oak Tree Lane Community Centre was completed in Spring 1980 after an eighteen month contract period which took in one of the most severe winters experienced during this century, in 1978/79, and a mining subsidence wave which crossed the site during the construction period.

2. The Centre, shown in sketch form in Fig.1, comprises a main hall, approximately 25 m x 18 m on plan and 10 m high, flanked by ancillary buildings which include a multi-purpose room, kitchen, bar, reception, changing rooms and boilerhouse and fuel store. The complex was designed by Mansfield District Council Architects Department and provides community facilities for the rapidly expanding Oak Tree Lane residential estate.

THE SITE

3. Trial holes and boreholes indicated that the site was overlain by fairly even thicknesses of 0.4 m of topsoil on 2.0 m of soft to firm sand above rocksand which was known to

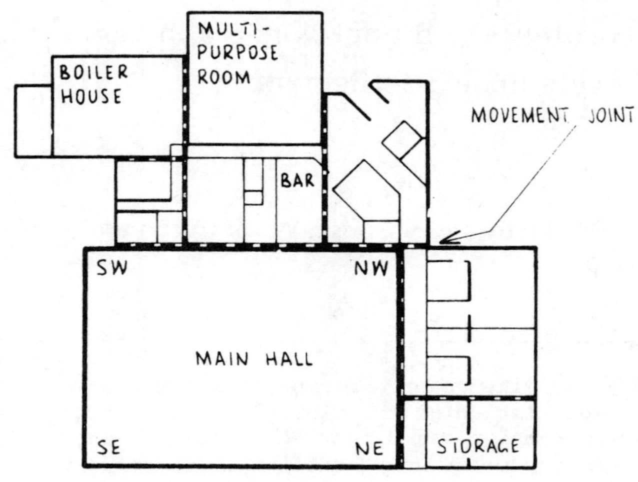

Fig. 1 Layout of building

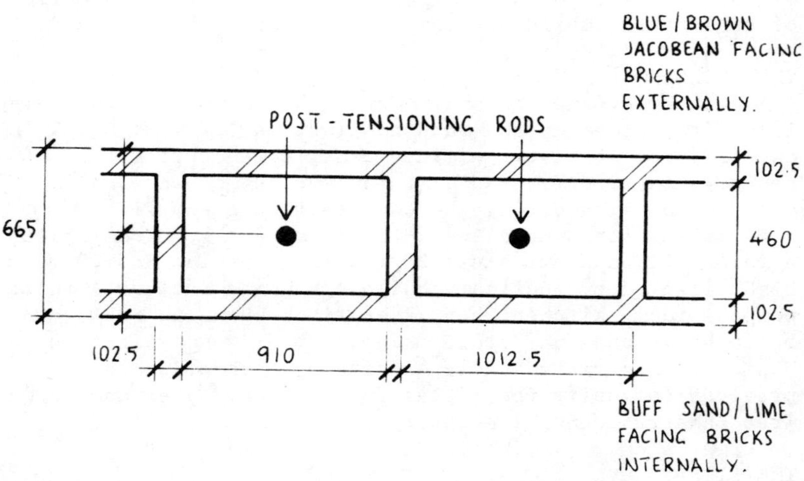

Fig. 2 Diaphragm wall section

be deeply fissured. The mineral valuers report confirmed that the site was situated in an active coal mining area with both past workings and future extraction to consider. The prospects of large ground strains, particularly tensile, through the fissured rocksand during the future mining subsidence waves demanded that the building load should be founded at as high a level as was practicable and that the foundation loads should be distributed as much as possible within the building plan.

THE MINING CONDITION

4. The NCB report indicated that, of the previous coal workings beneath the site, the most recent had been in 1973 and its subsidence effects had now ceased to effect the site. Three further seam workings were then being planned, the first and most significant of which was to commence during Spring 1979, with the remaining two operations to follow from 1981 onwards. The depths of the three seams planned to be worked were 400, 700 and 740 m respectively and the initial estimate of subsidence anticipated from the 1979 working was 700 mm. Detailed information of the effects of this subsidence, which appeared likely to affect the building during its construction period, was requested and the calculated results from the NCB are reproduced in Table 1 which gives a maximum subsidence, from total extraction, of 1080 mm and considerable differential subsidence across the site which would leave the building in a tilted state once the wave had passed through. The effects of the future workings, although considerably less severe, were predicted to have a righting effect on this tilt.

THE STRUCTURAL DESIGN

5. Loadbearing brickwork was selected as the most appropriate solution for the superstructure with diaphragm walls providing the vertical support to the main hall and traditional construction elsewhere. The diaphragm wall section adopted is shown in Fig.2. The ancillary buildings were all separated from the main hall with 75 mm wide movement joints which extended through foundations, superstructure and roofs and were introduced to accommodate the effects of the lateral and vertical movements of the subsidence. The ancillary buildings were also sub-divided with similar joints for the same purpose. The design of the main hall diaphragm walls would, in a non-mining situation, have followed the procedures and philosophy given in Structural Masonry Designers Manual (ref.1) and Brick

Table 1. Mining Subsidence Information

	Single Panel	Total Extraction
Maximum Subsidence	0.792m	1.080m
Permanent Strains (across panel)		
(a) extension	+ 1.52mm/m	+ 1.83mm/m
(b) compression	- 2.15mm/m	- 1.41mm/m
Travelling Strains		
(a) extension	+ 0.6mm/m	+ 1.0mm/m
(b) compression	- 1.0mm/m	- 1.75mm/m

The estimates of subsidence at each corner of the site (as shown on a plan supplied by Mansfield District Council - see copy enclosed) are as follows:-

	Single Panel	Total Extraction
North East Corner	768mm	994mm
North West Corner	792mm	972mm
South East Corner	546mm	1080mm
South West Corner	610mm	1069mm

	Single Panel	Total Extraction
Maximum Subsidence	792mm	1080mm

Estimated subsidence at each corner of the Main Hall (as shown on the enclosed plan) is as follows:-

North East Corner	728mm	1031mm
North West Corner	744mm	1021mm
South East Corner	649mm	1058mm
South West Corner	665mm	1053mm

Diaphragm Walls in Tall Single Storey Buildings (ref.2). Diaphragm walls have been used for such buildings for more than thirteen years and have proved themselves in both performance and economics. They are designed as a series of composite box or I sections to resist the lateral wind loading in bending. Unfortunately, there is no guidance given in the relatively new Code of Practice, BS5628, for the design of such walls although it is understood that attempts are being made to bring the code up-to-date to incorporate these and similar geometric forms. However, the design method described, in the publications refered to earlier, and used in the design of this structure, is based on sound engineering principles and the success of the existing buildings is evidence in itself of the suitability of that method.

6. The roof structure to the main hall comprises cranked steel beams which follow the pitched profile internally and provide an uninterupted soffit as can be seen in Fig.3 which shows the interior of the main hall. The main external elevations are shown in Figs.4 and 5. The roof beams were designed to span from wall to wall without a horizontal tie connection at eaves level and the minimal thrust on the top of the wall, due to deflection of the roof beams, was accommodated in the design of the diaphragm walls. The reinforced concrete capping beam at eaves level, which Contractors generally choose to precast, was, in this instance, poured in-situ. The inner leaf of the diaphragm was temporarily terminated 4 courses below the ring beam, during its construction, to facilitate the removal of the soffit shutters. The ring beam provided a positive bearing for the steel roof structure which was braced to afford support to the head of the diaphragm walls thus permitting them to be designed as propped cantilevers. The roof bracing system transferred the propping forces back to the stiff gable shear walls where, once again, the ring beam provided the positive anchorage.

7. The inclusion of the post-tensioning benefits, in the design of the diaphragm walls, resulted from an assessment of the effects of the mining subsidence wave on the structure and will be discussed in detail after a description of the foundation system adopted.

8. For the ancillary buildings, the plan areas had been sub-divided into relatively small units with total separation of each unit. Flexible raft foundations were designed which comprised 150 thick mesh-reinforced slabs tied into cage-reinforced edge beam thickenings of approximately

Fig. 3 Internal view of Main Hall

Fig. 4 External view

Fig. 5 External view

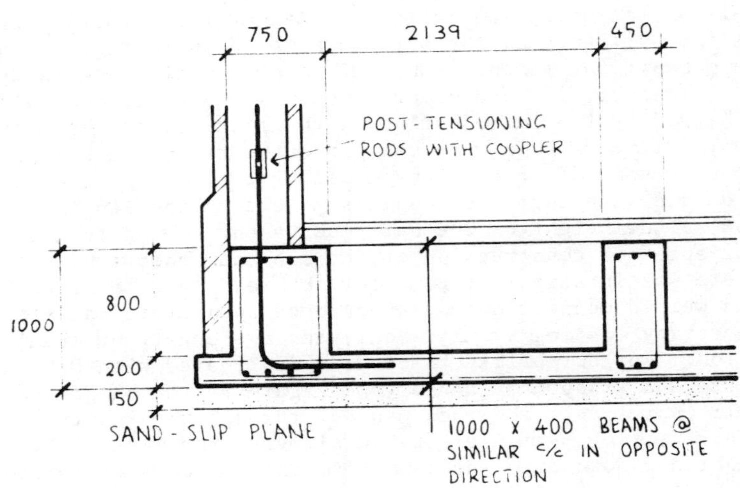

Fig. 6 Edge detail of raft foundation

600 mm square cross-section. In accepting the possibility of minor subsidence damage between adjacent units, due to differential movement, it was considered that the flexible raft solution provided adequate stiffness within each unit.

9. For the main hall, which could not be sub-divided into smaller units, a semi-rigid cellular raft was designed on the basis of one third of its length being capable of cantilevering and two thirds spanning. The cantilever and the span were each designed to carry the full dead and imposed loading from the building in the event that the subsidence wave should, at any one time, relieve that extent of support from the foundation. The raft comprised a 200 thick reinforced concrete base slab with a two-directional grid of upstand rc beams with an overall total depth of 1000 mm. There was no top slab to the raft as the floor covering to the main hall was to be sprung timber joist and hardwood strip flooring. A section through the edge of the cellular raft is shown in Fig.6.

THE POST-TENSIONING PHILOSOPHY

10. Consideration was given to the curvature of the cellular raft in its designed span and loading conditions and the effects of this curvature on the superstructure masonry was assessed. The deflected shape of the external wall of the hall, indicating in exaggerated form the effect of the curvature on the brickwork is shown in Fig.7. To compensate for masonry's notorious poor resistance to tensile stresses, induced compression was necessary and this was acheived by the introduction of the post-tensioning rods. 25 mm diameter rods were cast into the raft edge beams, one to each void of the diaphragm wall. The rods were to be provided with threaded couplers to extend the length of the rod as the height of the wall increased. Owing to the need for speed of construction of the sub and superstructure, there was insufficient time for threaded rods to be obtained and welded connections were approved subject to satisfactory integrity testing by ultrasonic and dye penetrant examination. For durability the rods were treated with Denso paste and wrapped in Denso tape, again as the height of the wall increased. The rods project through the insitu concrete ring beam, set in PVC sleeves, and were later tightened down onto the ring beam through a steel spreader plate to a specified torque using a simple torque wrench. The capping beam detail is shown in Fig.8. Alternate rods were torqued on the first circuit and the intermediates were completed on the second circuit and the whole of the

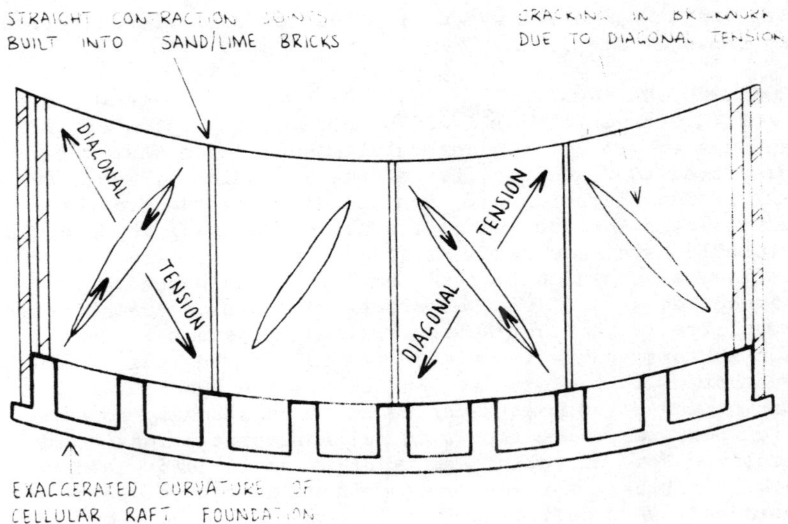

Fig. 7 Effect of raft curvature

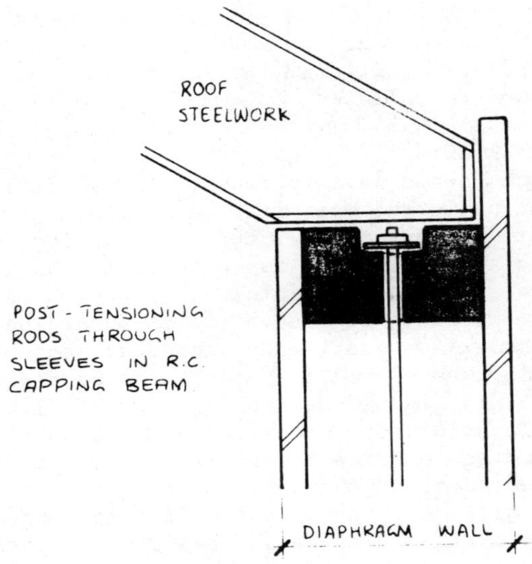

Fig. 8 Capping beam detail

tensioning operation was carried out by a single operative as shown in Fig.9.

CONSTRUCTION ON SITE

11. Construction commenced on site on 11 September 1978 and the effects of the next mining subsidence wave were predicted to reach the site at the beginning of February 1979. Consideration was given to delaying the project until after the subsidence effects had ceased, which would probably have incurred a delay of some two years, but other influences demanded that the building should be erected without delay and speed of erection became an essential pre requisite of the Contractors building programme. The Contractors task was not made any easier when the severe weather of that winter set in towards the end of 1978. Except under the most exceptional circumstances, construction work was maintained with suitable precautions being employed for concreting and bricklaying in cold weather. By mid-February the substructure and superstructure of the main hall were sufficiently well advanced to cope with the massive subsidence movements in the manner provided for in the design. The impressive speed of construction had been acheived by the Contractors diligence of effort, knowledgeable pre-planning and cooperation of employees. The quality of workmanship achieved at such high speed is a credit to the Contractor and provided yet further evidence against the misplaced criticism that masonry structures are inevitably slower to build.

THE SUBSIDENCE EFFECTS

12. As the subsidence wave approached the site, levelling points were set up in the main hall to monitor the extent of the movement. Unfortunately, the datumn point, which was established as far away as was practical, was still within the area of influence of the subsidence wave and consequently no accurate figures of total subsidence are available. The maximum subsidence recorded, below the displaced datumn, was 890 mm, and the total subsidence, estimated from this figure, was in the order of 1200 mm. The maximum tilt recorded, across the diagonals of the main hall, was 130 mm and occured around mid-July, some 5 months after the commencement of the subsidence effects on the building. Graphs were plotted, which are reproduced in Fig.10, to monitor the relative levels of the corners of the main hall. More recently, a further set of levels was taken which showed that the main hall floor has adopted a

Fig. 9 Simplicity of post-tensioning operation

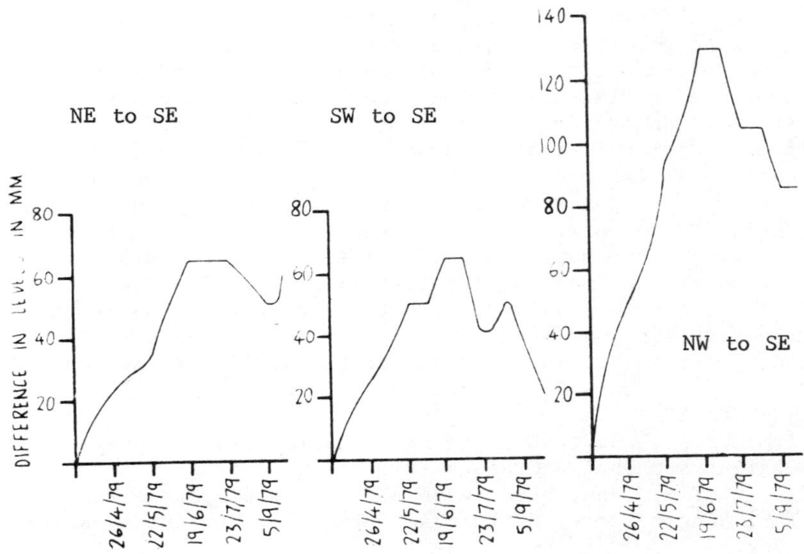

Fig. 10 Graphs monitoring subsidence of main hall

'permanent' tilt of approximately 55 mm. The future, and imminent, mining wave is predicted to correct that tilt by approximately 45 mm and consequently no relevelling of the suspended timber floor was undertaken, indeed it should never be necessary.

EXPERIENCE AND CONCLUSIONS

13. A thorough inspection of the structure was carried out by the design team in the company of N.C.B. Subsidence Engineers when it was considered that the effects of the subsidence had ceased. The inspection revealed that there was no cracking of the brickwork of the main hall attributable to the mining subsidence movements and the structural design was highly praised by the N.C.B. Engineers.

14. It is believed that this project is unique in that it provided the first ever use of post-tensioning applications to diaphragm walls and that to solve a mining subsidence problem. The technique was found by the Contractor to be extremely simple to effect and proved to be rapid in completion. The potential for it's application is far-reaching and is generating rapidly-growing interest through out the industry by way of enquiries to the Authors. Research work has recently been concluded, by W. G. Curtin and assistants at UMIST, into the behaviour of more heavily stressed post-tensioned diaphragm walls and the report and conclusions derived from this work are expected to be published during the current year. Mr Curtin believes there to be a considerable market for this technique, replacing the traditional steel or concrete framework, saving both costs and time on site and providing a more robust and maintenance-free structure. In todays society the vandal-resistance of brickwork is a benefit which cannot be overlooked and the social acceptability of the appearance of brickwork does not require justification. The Oak Tree Lane project provided valuable practical experience to add to the research findings mentioned earlier to provide a sound basis for future application and innovation of the technique.

REFERENCES

1. CURTIN, SHAW, BECK and BRAY. Structural Masonry Designers Manual. Granada Publishing, 1982.

2. CURTIN, SHAW, BECK and BRAY. Brick Diaphragm Walls in Tall Single Storey Buildings. Brick Development Association 1982.

12. Codification of reinforced and prestressed masonry design and construction

B. A. HASELTINE, BSc, ACGI, DIC, FICE, FIStructE, MConsE,
Jenkins & Potter, London

SYNOPSIS. Reinforced masonry has been used since 1825 and it has long been accepted that its design can be based on the principles used for reinforced concrete. There is a lack of detailed design guidance available in code form but the Draft British Standard Code BS 5628 : Part 2 has passed the Public Comment stage. It is discussed in detail in the paper.

INTRODUCTION

1. Masonry, meaning an assembly of bricks or blocks with mortar, has been used for centuries; the addition of reinforcement to strengthen plain masonry was first done by Brunel in 1825 on brickwork caissons to a tunnel under the Thames. Much of the reinforced masonry used this century has been in countries, such as the USA or New Zealand, where the need to resist seismic forces makes unreinforced masonry unattractive. On the other hand, in countries with no seismic problems, it is possible to build such highly stressed brickwork that there has been no incentive to add reinforcement. This has been particularly true in the U.K. where tall loadbearing brickwork buildings have been constructed without reinforcement [1.2].

HISTORICAL REVIEW

2. It was in 1943 that the first codified design information was published as "British Standard Specification for Reinforced Brickwork". Strangely this code for reinforced brickwork preceded one for unreinforced, although it dealt only with simple design and then only of walls.

3. A draft Code of Practice for the design of all loadbearing walls, including the relevant part of the 1943 Code, was circulated for comment in 1946. It had been

prepared for the Institution of Structural Engineers on behalf of the British Standards Institution (BSI) and included a general section on loadbearing walls, followed by sections on masonry, including brickwork(unreinforced), on reinforced masonry and on concrete cast insitu. After appropriate discussion and revision the document was published in 1948 as CP 111 "Structural Recommendations for Loadbearing Walls"(3). CP 111 used the then normal permissible stress approach, where the safety factors are hidden in the stresses given.

4. The first revision of this Code of Practice was published in 1964. The main change affecting brickwork and blockwork was an increase, usually substantial, in the permissible stresses; the basic stresses were altered slightly but the reduction factors for slenderness were made less onerous and were extended to include the effects of eccentric loads.

5. A further revision was published in 1970 as part of the programmed change in the construction industry from Imperial to SI units but, although there were other minor changes, the new code did not constitute a technical revision.

6. From the beginning, the code was based on the assumption that normal principles of structural design would be used to assess the loads resulting from a structure on its masonry elements. The detailed clauses of the code then gave guidance to enable wall thicknesses to be determined in relation to stresses that were considered to be safe and permissible, based on an assessment of experimental data and practical experience.

7. CP 111 contains a very short section on reinforced walls in Clause 320 which recognises that the design of reinforced brickwork and blockwork will be based on the same general principles of analysis as those used for reinforced concrete, referring to CP 114, the Concrete Code(4). Prestressed masonry is not mentioned and nor, of course, is CP 110(5), which was not available in 1970. A modular ratio approach to the design of reinforced walls is suggested; although the whole of CP 111 is stated to be devoted to the design of loadbearing walls, there is recognition of flexural stresses, so that there would be no difficulty in making use of CP 111 for other than reinforced walls, if the designer so wished.

8. Because there was so little guidance available on reinforced masonry a committee of the British Ceramic Research Association, the Structural Ceramics Advisory Group,

which consists of engineers,architects,contractors and quantity surveyors active in the field of masonry,together with brickmakers, commenced work on a Design Guide for Reinforced Brickwork in 1972 and it was published in 1977 as SP 91(6), in limit state philosophy.

9. Apart from CP 111, some information of use to designers has appeared in the form of technical papers and trade guidance, often reporting specific research projects. There are many documents available in the U.S.A where reinforced masonry has been used to a large extent.

10. A revised version of CP 111 was issued for public comment in 1974, and, after much discussion the limit state code BS 5628 : Part 1 : Unreinforced Masonry(7) was published in October 1978. Immediately afterwards, consideration was given to preparing Part 2 to deal with reinforced and prestressed masonry.

11. The Property Services Agency let a contract to the British Ceramic Research Association to write a draft suitable for the 'Public Comment' stage of BSI's procedure with a timetable of 1 year for drafting. Naturally, the basis for the draft was SP 91(6) although the considerable assistance of the Cement and Concrete Association enabled the purely brickwork parts to be expanded to include concrete block masonry. The draft was open for public comment from 1st May to 31st August 1981 and work in producing the final version of the code is now under way.

BS 5628 : PART 2 : REINFORCED AND PRESTRESSED MASONRY

12. The more important aspects of the draft code are discussed below.

Design Principles

13. The purpose of design is defined as the achievement of acceptable probabilities that the part of the structure being designed will not become unfit for the use for which it is required,i.e. that it will not reach a limit state.

14. Two limit states are recognised :

a) The ultimate limit state. In this an assessment using the design loads should ensure that an ultimate limit state is not reached as a result of rupture of one or more critical sections by overturning, buckling or twisting, caused by elastic or plastic instability. The layout on plan is required to ensure a robust and stable design,with some allowance for accidental loads.

b) The serviceability limit state. In this, it is required that deflection and cracking should not affect adversely the appearance or efficiency of the structure.

15. Loading. The loads to be applied should ideally have been determined statistically, but it is recognised that it is rarely possible to do this at the moment and so code loadings are required to be used,multiplied by a partial safety factory γ_f to give the design load. The values of γ_f are similar to those in BS 5628 : Part 1, i.e. for the ultimate limit state

Dead and imposed load acting
0.9 or 1.4 on dead load
1.6 on imposed load

Dead and wind load acting
0.9 or 1.4 on dead load
1.4 on wind load

Dead,imposed and wind load acting
1.2 on dead imposed and wind loads.

16. When considering overall stability,the design horizontal load should be not less than 1½% of the dead load above the level being considered.

17. Each combination of γ_f should be considered and the worst result used in the design. In practice, the worst combination is often obvious, and the work is what should have been but might not have been done using a permissible stress code. For the serviceability limit state γ_f is usually 1.0 or 0.8.

18. Materials. The design strength of the material is the characteristic strength divided by another partial safety factor γ_m which varies for brickwork in compression (γ_{mm}), brickwork in shear (γ_{mv}), steel (γ_{ms}) and for certain other uses. γ_{mm} may be varied according to the level of manufacturing control exercised in the production of the units i.e. for the ultimate limit state γ_{mm} is 2.5 or 2.8 for special or normal category of manufacturing control respectively. No variation is given for the degree of construction control, unlike Part 1, since it is stated in the draft that supervision must be up to the special

category of Part 1

19. γ_{mv} is 2.5, γ_{mb} (bond) is 1.5 and γ_{ms} is 1.15. For the serviceability limit state γ_{mm} is 1.5 and γ_{ms}, 1.0

Characteristic Strengths

20. The characteristic compressive strength of masonry (f_k) stressed perpendicular to the normal bed joints is the same as for unreinforced masonry, except that the range of mortars permitted is restricted to the two strongest designations, (i) and (ii).

21. When solid or frogged bricks are stressed in other directions the same figures are permitted; for perforated bricks only 0.4 of the tabulated strengths may be used when the units are stressed other than perpendicular to the normal bed joints and for unfilled hollow blocks, tests must be used to determine the compressive strength for the particular direction of loading.

22. SP 91 permitted the bending compression to be taken as 1.5 times the characterstic compressive strength; in the draft code this factor has been reduced to 1.2 but it has been the subject of considerable debate and may need to be modified. The problem is that there have been very few tests resulting in pure compressive failure and so the correct factor is not readily available. In practice it may not be significant, as reinforced masonry sections will rarely be over-reinforced.

23. Shear seems always to present a problem and it has in Part 2. For sections in which the reinforcement is placed in mortar (i.e. bed joints or small cavities formed by the bond of the units) the characteristic shear strength is 0.35 N/mm^2. Where the steel will be surrounded by concrete the values are

$100 \frac{As}{bd}$	Up to 0.15	0.5	1.0	1.5	2.0
f_v	0.35	0.45	0.55	0.60	0.65

24. Racking shear is given as in Part 1 i.e. (0.35 + 0.6 x design vertical load) N/mm^2.

25. The strength of reinforcement and prestressing strand are based on those in CP 110.

Axial Compression

26. The strength of reinforced members in compression is treated in a similar way to unreinforced in Part 1; a slenderness ratio reduction factor approach is used, but the additional strength attributable to the steel is allowed for.

Bending

27. The bending strength of reinforced masonry is based on the following assumptions :

a) Plane sections remain plane.

b) The stress distribution in the masonry is rectangular with an intensity of the characteristic bending compressive strength divided by γ_{mm}.

c) The maximum strain in the compression zone is 0.0035.

d) The depth of the compression stress block does not exceed half the effective depth.

28. The resulting formula for the Moment of resistance of the masonry is

$$\frac{0.375\ f_f\ bd^2}{\gamma_{mm}}$$

where f_f is the bending compressive strength (see 22).

29. Deflection and cracking must not be such as to cause serviceability problems, but it has to be admitted that the means of calculating these two items are not very satisfactory.

Combined Bending and Compression

30. Columns and walls are divided into those that are "short",i.e. where buckling is unlikely and those that are "slender" where second order effects may be important.

31. In the former case no additional bending moment due to buckling is to be allowed for, whereas in the latter, second order affects are included in the formulae given.

Prestressed Masonry

32. The general principles of prestressed concrete design are used and only supplemented as necessary for masonry. In the absence of better information creep is assumed to be numerically equal to twice the elastic deformation under the action of the prestressing force.

Corrosion Protection

33. Probably the aspect of reinforced or prestressed masonry that receives most discussion, at present, is the protection of the reinforcement to prevent corrosion. It seems to be agreed that where reinforcement may be contained in mortar or concrete that will carbonate and then be subjected to wetting and drying cycles, so that fresh oxygen can reach the steel, then only heavily

protected steel, or stainless seel, should be used[8,9,10]. Once that decision is made, the cover for exposed reinforced masonry becomes a matter only for bond and structural considerations.

34. It appears that mortars carbonate quickly if they are porous, or are between gas porous bricks. Concrete and dense mortars, especially when used with high strength bricks, carbonate slowly, and corrosion of steel in those cases is less likely.

Detailing and work on site

35. The minimum percentage of reinforcement to be used in reinforced masonry is given as is advice on the detailing of the bars.

36. A section on Work on Site, dealing with materials, workmanship and quality control, completes the draft.

REFERENCES

1. JENKINS,R.S., and HASELTINE, B.A., Recent Experiences with Calculated Loadbearing Brickwork,British Ceramic Society Proceedings No.11 July 1968.

2. HASELTINE, B.A., and AU, Y.T., Design and Construction of a Nineteen Storey Loadbearing Brick Building, British Ceramic Research Association,Proceedings of Second International Brick Masonry Conference.

3. BRITISH STANDARDS INSTITUTION,The Structural Recommendations for Loadbearing Walls. CP 111 : 1970.

4. BRITISH STANDARDS INSTITUTION,The Structural Use of Reinforced Concrete in Buildings. CP 114 : 1965.

5. BRITISH STANDARDS INSTITUTION,The Structural Use of Concrete. CP 110 : 1972

6. BRITISH CERAMIC RESEARCH ASSOCIATION,Design Guide for Reinforced and Prestressed Clay Brickwork,B.Ceram. R.A.Spec.Pub.91 1977.

7. BRITISH STANDARDS INSTITUTION,Structural Use of Masonry BS 5628 : Part 1 : 1978.

8. DE VEKEY, Dr.R, Durability of Reinforced Masonry, Institution of Structural Engineers,Reinforced and Prestressed Masonry July 1981.

9. FOSTER, D.,THOMAS,A.,Aspects of Durability of Clay Brickwork,Institution of Structural Engineers, Reinforced and Prestressed Masonry July 1981.

10. ROBERTS, Dr.J.J.,Interim Results from an investigation of the durability of Reinforcing Steel in Reinforced Concrete Blockwork,Reinfored and Prestressed Masonry July 1981.

Discussion

PAPERS 1 & 2

Dr G.J. Edgell (British Ceramic Research Association, Stoke-on-Trent)

1. It is good to see the Institution of Civil Engineers acting as one of the sponsors for this meeting. I hope that this is a sign that the Institution is taking a greater interest in structural masonry than has recently been the case and that this interest will be sustained by its new Structural Engineering Group.

2. My own activity is principally research, which is largely funded by the brick industry. I am also involved with the drafting of the new code of practice on reinforced and prestressed masonry.

3. Dr Curtin questions the value given in the draft code for the modulus of elasticity for brickwork, which is 900 x characteristic strength. Although there will be some variation in the accuracy of this estimate for different types of brickwork and different structural elements it seems to be of the correct order. From the results of experimental determinations on piers, made at B C R A and elsewhere (Fig.1), it seems that the draft code estimate is reasonable, although possibly tending to underestimate the stiffness for the higher strength brickwork. Although it is not always wise, particularly when considering material properties, to look at overseas practice, Table 1 shows values that have been given as guidance in some other countries. Clearly the choice in the draft code is reasonably consistent with these. The first British Standard on reinforced brickwork, BS 1146 : 1943, gave values for the modular ratio for various brick strengths, and these are also reasonably consistent with the draft code.

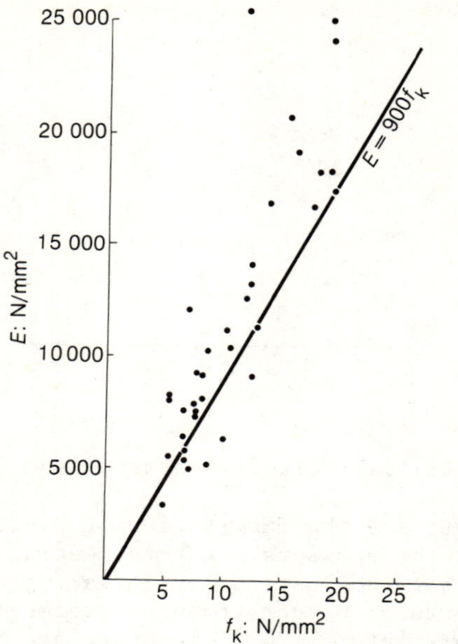

Fig.1. Relationship between E and f_k

Table 1. Modulus of elasticity values

Structural Clay Products Institute (USA)	$1000f_m$
New Zealand	$1000f_m$
Australia	$750f_m$
Canada	$1000f_m$
Russia	$500f_m$, $750f_m$ or $1000f_m$ (depends on mortar)

4. Although I believe that the E value given is reasonably accurate, Dr Curtin's concern suggests to me that the code ought to point out that the engineer may wish to determine E by test rather than by use of the value given.

5. Another point, although minor in terms of the broad picture set out by the paper, but significant from my point of view, is the indication that the pocket type wall is an established technique for low walls only. The brick industry's research programme has been on 3 m high walls, and as is pointed out in Mr Sutherland's paper the tallest wall of this type was reported to be 7.3 m in 1971. It is important that pocket type walls should be considered as a possible method of retaining a large height of soil.

6. Dr Curtin makes the point that we need be careful about corrosion of steel, and pictures of a cracked brickwork building where the reinforcement appears to have had no mortar or grout about the bars give the reminder of what can go wrong. However, this is not the only side of the coin. Recently I was sent details of a bridge in Ohio where a supporting pier of dense good quality brickwork containing steel one brick thickness from its face was still sound after 49 years, and that despite the water level fluctuating about it. Similarly, reinforced brickwork silos are still is use near Chicago after 51 years during which there have been no corrosion problems. Another example of how reinforced brickwork can be durable is a compressor building near St Louis which was built in 1932 and eventually dismantled in 1960, again with no record of any problems due to corrosion. The reinforced brickwork proved to be extremely strong when a tornado swept through the chemical plant.

7. Clearly corrosion is not always a problem, but the code-drafters are aware of the need for caution. If anybody knows of any reinforced masonry that could be examined for evidence of corrosion, I would be very pleased to hear from them.

Dr E.W. Bennett (University of Leeds)

8. In the tests of reinforced brickwork beams referred to in Paper 2, two of the eleven beams were based on members in the actual structure and were tested to confirm the behaviour of a reinforced brickwork beam with a concrete topping (e.g., part of a floor slab) under positive and negative bending. These failed in flexure at greater loads than indicated by calculation.

9. The remaining beams were tested in shear to verify the design recommendations of SP 91 with varying amounts of stirrup reinforcement, in view of the limited amount of

experimental data available to the authors of this publication. The results of these tests showed the SP 91 values to be conservative, whereas the values (unfactored) given later in the draft BS code were about equal to the experimental shear strengths.

10. The tests underlined the importance of adequate ties connecting the brickwork with the grout core when the latter was large, since in the absence of ties some of the beams failed by separation at the interface. Except at high loads both flexural and shear cracking occurred by opening of the mortar joints in the brickwork and was not conspicuous. At service loads the deflexion of the beams was generally about equal to the value calculated by the elastic theory using the modulus of elasticity of brickwork according to SP 91.

Dr R.C. de Vekey (Building Research Establishment, Watford)

11. I note that in the office block described in Paper 2 there was a 'cocktail' of different metals being incorporated within the walls, including normal high yield steel reinforcement, Macalloy bars, stainless steel ties, stainless steel reinforcement and an assortment of loops, triangles, stirrups etc. Were any specific design precautions taken to ensure that the various different alloys were not touching each other in the finished building to avoid the problem of bimetallic corrosion? This would be particularly important in external walls where they would be expected to be damp or wet for parts of their life. Was there any quality control of the site specifically to avoid this problem?

Mr R.E. Watson (School of Architecture, Hull College of Higher Education)

12. I can readily appreciate the use of blockwork as a structural material, but presume that when a fair finish is required in brickwork, cost will be a significant factor.

13. On visiting the works and offices of George Armitage & Sons Ltd, I was very impressed with the structural applications, and noted the care taken in achieving the finish. I should like to ask Mr Bradshaw and Mr Drinkwater if they can give an indication of cost.

14. One apprehension I have concerns quality control and site supervision. One of the papers mentions developments in the Third World, where supervision is likely to be minimal - especially so if self-help schemes on small budgets are contemplated and where local materials may be of low quality. Not only does the professional engineer have a responsibility, but also technicians who essentially will be carrying out the controls, but of whom there may be a lack. I would ask

speakers to comment on the requirements and necessity of adequate supervision.

Mr C. Southcombe (Plymouth Polytechnic)

15. Dr Curtin gives as a crude guide (Paper 1, paragraph 4.0) the preference to prestress brickwork and to reinforce blockwork. This advice ignores particular situations where reinforced brickwork will be used (e.g., reinforced brickwork beams subjected to vertical loads, and walls reinforced in their bed joints resisting lateral loads). There are many builders/civil engineering contractors who would not have the confidence to use prestressing without good supervision, which many are unable to provide.

16. A beam using a quetta bond type brickwork construction can be cheaper than a reinforced concrete beam of similar strength. Tests on quetta bond beams of 3-6 m span at Plymouth are leading to proposals that

(a) serviceability design is generally much more significant than strength design;
(b) within the serviceability load range the tensile strength of brickwork is significant enough not to be ignored;
(c) comparing reinforced brickwork with reinforced concrete (paragraph 5.3) is not the way forward; reinforced brickwork design requires a 'new approach'.

Dr Curtin (Paper 1)

17. In reply to Dr Edgell, I have doubts that a blanket cover of an E value of $900f_k$ is adequate. Although this value has been obtained from different laboratories it is possible that they have carried out identical tests on identical specimens. Site experience has shown variation of E value depending on type of brickwork, the height and shape of the element, the method of loading, compressive and bending stress, etc. I would suggest that the code should not be too dogmatic about the E value and fully agree with Dr Edgell that it should suggest that the engineer might prefer to carry out his own tests on his own site.

18. Where the paper indicates that the pocket type retaining wall is an established technique for low retaining walls, this means up to 3 m or so. Certainly they can be built taller but site-costing has shown, in the past, a decreasing cost-effectiveness with further increase in height compared with other techniques. It may be worthwhile carrying out case cost-studies on site to obtain reliable comparative data - it may be a fruitless exercise doing a

'desk' study (which can 'prove' anything!)

19. I do not think we can afford to be sanguine about corrosion. The Ohio bridge with its one-brick cover of 'dense, good quality brickwork' will probably last for years. It is possible that the Chicago silo and the St Louis compressor building had equally excellent cover. Corrosion of reinforcement in masonry is not a problem if there is adequate cover - neither is it in reinforced concrete. The problem is the provision of adequate cover by fully continuous, thoroughly compacted dense grout of the correct mix, workability, depth etc.

20. In reply to Mr Watson, the apprehension concerning quality control and site supervision is very understandable. But 'the requirements and necessity of adequate supervision' naturally applies to all structures, in all materials, assembled under all conditions by all types of labour. The importance of the structure, the reliability of the materials, the care and skill of the construction, etc. can, of course, be allowed for in the adjustment of the partial safety factors. With good materials, design, specification, construction etc. the global factor can be reduced to 2 - with unfavourable conditions it could be increased to 6 or more.

21. Masonry construction is likely to be even more successful for the Third World since they are likely to have indigenous labour more experienced in using well tried local materials than, say, using structural steelwork or high grade prestressed concrete.

22. Taking up the points raised by Mr Southcombe, unfortunately in a brief but wide-ranging paper it was not possible to give more than a 'crude' guide (perhaps better described as 'condensed' or 'simplified'). Nevertheless, considerable practical experience of actual projects subject to competitive tender and cost control show the guidance to be correct - and this includes horizontal beams subject to vertical load, vertical walls subject to horizontal load and any other structural brickwork element subject to bending.

23. Prestressed brickwork is not the same high technology process as heavily prestressed concrete. Numerous projects have been successfully completed by small builders who have found no difficulties and suffered no qualms.

24. I am somewhat surprised at the statement that a quetta bond brickwork beam can be cheaper than a reinforced concrete beam and it would be interesting to see site cost records.

25. I would agree that 'serviceability design is more

significant than [ultimate] strength design' and that within the serviceability range 'the tensile strength of brickwork is significant'. However, the tensile strength can be difficult to predict and can be unreliable in general practice so it may be prudent to discount it in design.

26. I fully agree that 'reinforced brickwork design requires a new approach', and did not mean to imply that it was the same as for concrete. I was implying that there may be some researchers (more interested in producing their norm of papers than providing useful information) who may merely 'apply' concrete research to masonry research.

Mr Bradshaw and Mr Drinkwater (Paper 2)

27. In answer to Dr de Vekey the simple answer is that yes, efforts were made to avoid the problems of bimetallic corrosion and these were recognized at the design stage. In all positions where the reinforced brickwork was exposed to the weather, stainless steel was used throughout. In all positions where the reinforced brickwork was not exposed, high yield steel was used. Inevitably there were positions where the two met at changeover points but these were either inside the building or contained within reinforced concrete elements where sufficient concrete cover could be provided. The building was constructed during the steel strike and choice of reinforcement was limited.

28. Replying to Mr Watson; during the period of the contract, building costs escalated due to inflation (tender base month July 1979 and practical completion May 1981) and the NEDO fluctuations increased the final cost considerably. The final cost amounts to approximately £545/m^2. The client's requirement for adaptable working areas has been met by the reinforced brickwork structure of the two wings at a cost similar to typical high quality office building costs. The cost of the reinforced brickwork beams in the central area is similar to that for reinforced concrete beams faced with brick slips.

29. Supervision is clearly required for reinforced masonry as it is for other materials. Training is needed since site operatives must be convinced of the need to carry out the work in a particular way. Otherwise unacceptable work may result as soon as the supervisor's back is turned.

30. The construction of reinforced masonry in the Third World should be no greater problem than using other materials, and concern regarding workmanship and quality of materials can be met by increased factors of safety and ensuring robust structures.

DISCUSSION

PAPERS 3-5

Mr Sutherland (Paper 4)

31. Cost studies have been made on solid walls both with one-way bending (e.g., retaining walls) and two-way bending ('wind' walls); they are described in ref.7 of my paper. For both types of bending, the thickness of masonry needed for the same resistance moment is very similar either with reinforcement or prestressing. However, in the prestressed case substantially more 'steel force' is needed to maintain the 'no tension' serviceability state than in the reinforced case where tension and cracking are both permitted. The supply cost of prestressing rods and high yield reinforcement appears to be very similar per kilonewton force per metre length. Thus with a greater 'steel force' the supply cost of the steel must be greater with prestressing and if the masonry is unchanged the only scope for economy with prestressing lies in the relative cost of fixing, stressing and protecting the prestressing tendons (including anchor plates, etc.) compared with fixing and protecting the reinforcement. With prestressing it is difficult to see any real saving here - just the opposite in fact.

32. Using cellular (diaphragm) walls of practical form and scale the steel force still appears to be greater with prestressing than with reinforcement but here direct cost comparisons become more complex. It is not certain how beneficial the cellular form is in itself once it gets beyond the stage where it can act as a gravity structure.

33. The arguments above are based on cost alone. If the higher performance of a prestressed wall is needed there is a good case for prestressing. There may also be a case for encouraging partial prestressing, especially when considering wind forces.

Mr B.J.B. Gauld (School of Architecture, Kingston Polytechnic)

34. Due to the earthquake requirements in New Zealand, block partition walls have to be reinforced, both vertically and horizontally. Therefore concrete blocks reinforced are extensively used. Fig. 2 shows a partition wall under construction in the School of Architecture, Auckland. The blocks were carefully laid to produce a fair face finish both sides. The wall is connected to the slab but a movement joint is maintained between the columns and the soffit to allow the structural frame free movement during an earthquake.

35. Some studio houses in Highgate, London (Fig.3) were built from double-leafed 90 mm concrete blocks, fair-faced

Fig.2. School of Architecture, Auckland: concrete blockwork

Fig.3. Studio houses, Highgate: Forticrete concrete blocks

Fig.4. Block retaining wall

inside and out. At one end of the site, a retaining wall was required. This wall (Fig. 4) was constructed by a small builder, producing a first class finish and without the use of any shuttering.

Dr J.G.M. Wood (Mott, Hay & Anderson, Croydon)

36. Reinforced blockwork or brickwork has, and will continue to have, extensive application for retaining heaps of dry stored materials in both agriculture and industry. However, when silage is to be retained, as in Mr Adams' example, particular care is required to prevent corrosion of reinforcement. Silage often produces significant volumes of acid (acetic, lactic and/or butyric) with a high pore pressure against the wall. This will rapidly penetrate a

porous wall, decaying the mortar and corroding the reinforcement. Could Mr Adams explain what precautions he took to minimize this deterioration.

37. Mr Beard's circular brick bin and the measurements of floor loads are most interesting. His simple design method, discounting the effects of unloading and the eccentricity of discharge, is reasonable for a small, relatively squat bin. However, for larger structures these effects must be considered.

38. The discrepancy between Mr Beard's measured pressures and those calculated is likely to have arisen from variation of the actual density and friction properties from the approximate and varied values given in most references. Friction properties are particularly sensitive to grain moisture content. The approximate treatment of the conical central surcharge of grain in calculation may also have caused some of the discrepancy. Could Mr Beard explain the basis of measuring 'height of grain' in his Fig.4; is it to the top of the cone? What were the density and friction properties used for the calculation of the 'design' curve?

Dr J.J. Roberts (Cement and Concrete Association, Slough)

39. Where orthogonal walls of silage-clamp meet one can either design for movement or detail the reinforcement so that common structural action results. Both can give problems. I should like to ask what is Mr Adams' experience.

40. Some of the sections shown in Paper 3 are similar to columns. Has Mr Adams any views on the need for stirrups in reinforced concrete masonry? The steel percentage is usually low and the cover large - except in the case of stainless steel. Buckling could therefore be considered to be restrained by the masonry.

41. With reference to Dr Wood's contribution, the Cement and Concrete Association's publications for silage walls include a very detailed specification for the protective measures necessary for the masonry.

Mr G. Shaw (W.G. Curtin & Partners, London)

42. In Mr Sutherland's costing comparisons (paragraphs 31-33) of post-tensioned masonry as against reinforced, his costs must not have taken account of the simplicity of construction of the post-tensioned wall. The main economy is that the bricklayers can just build brickwork and forget about reinforcement details, grouting and post-tensioning. They merely build the wall with large voids which contain the post-tension rods; all the work of fixing the rods is carried out at foundation stage and all the post-tensioning is done

at roof level.

Dr D.P. Wyatt (Brighton Polytechnic)

43. During design, consideration must be given to the problems in putting the brickwork together and its subsequent satisfactory performance. The structural engineer may execute a sound calculation design but the specifier and architect may not present adequate design and construction details to site. Further, there are those real difficulties on some contracts which militate against sound quality control. Against this background, could Mr Sutherland expand on the risk of failure of retaining walls in brickwork. This request is prompted by the widespread problems of disintegrating cappings, omission of weepers (weep holes) and cracking. In the slides shown by Mr Sutherland I noted no weeper provision and no backwall drainage and it was not possible to identify the capping detail. Failure in these areas may militate against economic use of less expensive reinforcement.

Mr K.F. Tune (Ove Arup & Partners, Sheffield)

44. Would the authors of papers 3 and 5 please comment on whether they have experienced particular practical difficulties on site, bearing in mind that most of the jobs described are too small to have full-time supervision from either a Clerk of Works or a Resident Engineer.

Mr Adams (Paper 3)

45. Dr Wood is correct in calling attention to silage effluent as a corrosive and contaminating fluid, the control of which is an important part of any silage bay design. My own thinking on this has not changed, in essence, over the past 15-20 years. First it is necessary to provide drainage so that effluent can be got away from the back of the wall and discharged so that it cannot contaminate watercourses. Secondly, some protection should be given to the blockwork to control penetration into the wall while the initial pore pressures are dissipating. In the early days we thought that renders were necessary, then changed to bitumen coatings (which farmers tend to keep by them) and, latterly, two-coat chlorinated rubber paint (Cement and Concrete Association (ref. 1)). The main ingredients for applied coatings to be relevant would seem to be

(a) a dense flush pointed face to work on;
(b) sealing of that face with a cement-sand slurry;
(c) applying the coating down the wall and a short way on to the slab to cover the vulnerable internal corner.

I have seen unprotected walls reported to be over 12 years old, and still functioning; and the condition of many steel silage clamps I have inspected makes me suspect that silage effluent is not quite as corrosive as some fear. Nevertheless, I would not wish to omit this sensible protection on any clamp designed by my practice, particularly knowing that small and medium size farms have no tradition of tight control over building standards and that 'maintenance' and 'design life' are words remembered only when things stop working.

46. Dr Roberts brings up the interesting question of whether to build an open-ended box or two freestanding cantilevers with another cantilever across the end. The proportions of silage clamps do not give rise to any structural savings from providing continuity on plan. Much of the wall acts as a free cantilever anyway. I am in favour, instinctively, of keeping things simple and sticking to three straight wall lengths of identical section and reinforcement. One problem with this approach is theoretical loss of effluent into and through the joint. I do not think this is as serious as it first appears because the corners are the most difficult parts to compact the silage into, and pore pressures dissipate quickly here to the drain. A simple butt joint against a bituminous paint face to break the restraint I would think adequate in most cases. I have used continuity round the corners but this has probably not shown any advantage. Moisture movements and stresses from the loads are not easily catered for via bond on L bars.

47. Concerning Dr Roberts' second point, in the examples shown in the paper I do not think that links are necessary. The proportions used in these and most practical examples would suggest to me that links may be omitted and that buckling will not be a significant factor or will be prevented by adjoining masonry.

48. Dr Wood is correct in implying that more slender, isolated columns with higher percentages of vertical reinforcement may need to be treated differently. Tests have been made on slender columns (200 mm square in section) reinforced centrally with a 25 mm dia. high yield bar, in order to check various theories for predicting ultimate load, under various axial loads and moments, against reality. It was noticeable that tests under axial load only gave failure earlier than was predicted and there was reason to suspect this was initiated by buckling of the reinforcement.

49. Mr Tune's assumption that most of the jobs were too small to have full time supervision is not correct. Most do.

50. Reinforced blockwork is not a cure-all and the examples

given are of elements, within a wide range of job types and sizes, where reinforcing blockwork was a valid answer for that element.

51. That aside, the answer to Mr Tune's question would be 'Yes', and the specific case of a silage wall is almost the supreme example. There will always be difficulty in enforcing standards where the client does not wish to pay for adequate supervision, whatever material is used. The engineer must learn to overcome this. He must define his service, specify clearly everything that is critical, emphasize it in layman's language together with the likely implications of skimping, and advise strongly on the minimum supervision necessary. In many cases he must design to suit the standard of workmanship he is going to get.

52. The hardest job is to convince a farmer that reinforced brickwork is not something he can competently tackle himself between harvesting and ploughing. Blockwork still suffers from the stigma of being a cheap, replacement material for the rough job; the breeze block image. Engineering blockwork is a long way ahead of the layman's conception of it.

Mr Sutherland (Paper 4)

53. The cost comparisons which Mr Shaw questions were not based on estimated contract prices or tender prices adjusted to suit an elusive level of comparability. They were the logical, but at first slightly surprising, outcome of a comparative structural performance study between reinforced and prestressed masonry walls based first on qualities of materials used and then considering the supply cost of these and finally looking at costs on site.

54. Dr Wyatt and Mr Tune both ask questions on workmanship and supervision in relation to reinforced brickwork. There is no simple answer here especially on small projects. However, brickwork is a very forgiving material and provided that the steel is reasonably protected other errors of workmanship may well never come to light! Stainless steel (or stainless clad steel) provides a virtually complete safeguard against corrosion and where quantities needed are small the overall extra cost could be well worth while.

55. Dr Wyatt also questions the apparent lack of weepholes in the retaining walls I showed. These were provided as indicated in Fig. 1 of my paper.

Mr Beard (Paper 5)

56. I am grateful for Dr Wood's comments on the acceptability of the simple design method for the size of silo described in my paper. With regard to his questions, the

height refers to the mean height related to the volume of barley in the silo. I do appreciate, however, that this contributes to a discrepancy between measured and calculated pressures, but the error diminishes as the height of grain increases. The density and friction properties used in the calculation of the design curve were 800 kg/m^3 and 0.4 respectively, these being the values assumed for wheat.

57. With regard to Mr Tune's question, there were no practical difficulties in building the circular silos or the rectangular silo. I would add, however, that the groove in the curved block used in the circular silos (Fig.3 of the paper) was essential to hold the hoop reinforcement in the correct place. In the quetta bond wall used for the rectangular silo the vertical reinforcement had to be braced to ensure that it was correctly located. Using a bar in alternate pockets was also important because it permitted the bricklayer to work from one side and lay bricks through the reinforcement.

PAPERS 6-9

Dr D.P. Wyatt (Brighton Polytechnic)

58. Mr Johnson's slides showed shear reinforcement built into the brickwork, together with wall ties. The use of wall ties puzzles me and I wonder if Mr Johnson could comment in view of the lack of research work on the performance of wall ties in cavity work. What design value, if any, was taken; and more to the point could he explain what value the wall ties serve? It is difficult to see from a construction point of view whether they are necessary even when 'pugging' the wall in the four-course lifts mentioned.

Dr R.C. de Vekey (Building Research Establishment, Watford)

59. In the structure Dr Curtin described (Paper 8) I noted that the concrete ring beam appeared to bear only on the inner leaf and cross-webs. Was this a correct impression? If it was, would not the precompression and differential shrinkage induce a shear stress at the junction of the webs and the outer leaf and is there no danger of a shear failure here?

Mr D.R. Astin (West Yorkshire Metropolitan County Council, Wakefield)

60. I should like to ask Dr Curtin how the stability of a diaphragm wall is provided during construction prior to the

roof deck providing support.

61. Mr Bradshaw (Paper 7) has described the composite action between reinforced concrete ground beams and the supported brickwork. Was a damp-proof membrane provided in this situation?

Mr S.B. Zukas (ZMCK Consulting Engineers, London)

62. For the church described in Paper 7, Mr Bradshaw has referred to composite design of the walls spanning over piers and has referred to a graph showing the arching effect (ref. 1 of paper). There has always been some difficulty in applying Dr Wood's work at BRS on composite action when dealing with continuous composite panels. How did Mr Bradshaw deal with continuity over the supports?

Mr P.R.G. Woolley (Hannah, Reed and Associates, Cambridge)

63. Would the authors comment on measures thought necessary to prevent corrosion of reinforcement. I would feel more confident in proposing the use of reinforced masonry if I could be sure that the extent of the problem was known and that adequate practical preventive measures existed. There are two cases to consider. One is bed joint reinforcement, which I understand is more prone to corrosion. Do the authors consider galvanizing to be an appropriate protection, or do they invariably recommend stainless steel? The other case for consideration is protection of vertical reinforcement in hollow infilled blockwork, particularly the practical difficulty in achieving on site the design cover.

Mr A.S. Safier (Armand Safier & Partners, Stanmore)

64. The emphasis that has been given by nearly all the authors, and other contributors, to the risk of corrosion to the reinforcement in reinforced masonry makes me think that the symposium would like to know about a development in this field.

65. For about 5 years, within our work as consultants, I have been advising a specialist firm (Coated Reinforcement Ltd, Luton) on the use of electrostatically epoxy-coated reinforcement and researched its applications.

66. Electrostatic coating of reinforcement was developed by the USA Department of Transportation's Federal Highway Authority (FHA) starting some 12 years ago as an answer to the very severe corrosion incurred on their bridges due to the application of deicing salts. It has by now been fully proven and tested by the FHA and is produced to comply with a new code, ASTM A775-81 (May 1981). Many thousands of

tonnes of such reinforcement have by now been used, including about 40 t in a bridge project by the Nottinghamshire County Surveyor. Indeed, in the USA, Federal funding for highway bridge construction is only provided if epoxy-coated reinforcement is used in the project.

67. Epoxy-coated reinforcement appears to be ideally suited for use in reinforced masonry and is now being made available in the UK by Coated Reinforcement Ltd. In the manufacturing process the coating powders have to be specifically formulated and applied by the electrostatic method on purpose-built equipment.

68. I have suggested to BSI Committee CSB/33, which is producing the new reinforced masonry code, to allow within the code for the use of epoxy-coated reinforcement. The method has in my opinion advantages over the use of metallic coatings such as zinc and stainless steel, avoiding bi-metallic corrosion problems and being substantially more economical.

Dr R.C. de Vekey (Building Research Establishment, Watford)

69. Taking up a point raised by Mr Safier, BRE is in the process of assessing the durability and resistance to mechanical damage (scuffing) of powder-epoxy-coated reinforcement but the tests have not been of sufficient duration to enable comment to be made on the durability of this material as yet.

Dr G.J. Edgell (British Ceramic Research Association, Stoke-on-Trent)

70. I am acting as Chairman of a small BSI working group studying the durability of reinforced masonry in relation to the public comments that were made on the draft code of practice.

71. The original draft was based on the philosophy that the masonry provided no protection to the steel. This arose from a preoccupation with carbonation of mortar or grout and the fear that once this had taken place corrosion would begin and the structure would be rendered structurally unsound after a short time. However, after further investigation and an examination of the situation in reinforced concrete, and with help from the Cement and Concrete Association, it became clear that corrosion would not necessarily take place immediately, if at all, and that the structure could still have significant life. Also, one of the problems with the original draft was that because of the insistence for cover to be similar to that for reinforced concrete about the bars some forms of construction that had

been successfully used in reinforced masonry became impossible without the use of stainless steel reinforcement. The current draft now recommends, in, for example, grouted cavity construction, 20 mm cover about the bars, with different types of bar used in different conditions of exposure.

72. The situation with bed joint reinforcement is very variable. I have seen examples of galvanized bed joint reinforcement in a relatively sheltered situation that has rusted through within 10 years. Also there is unprotected reinforcement in a very large factory wall, in some cases with little cover as it was provided in coil rather than straight lengths and was difficult to place, which is in good condition. As the volume of the corrosion products is seven times that of the original steel, apparently rusty reinforcement may have lost little strength. The draft code is likely to indicate that galvinized bed joint reinforcement with a thickness of zinc to the new wall tie specification is adequate in most external situations. In sheltered situations unprotected steel may be used and in severe exposure situations, the exposure being defined on the basis of the driving rain index as in CP 121, stainless steel should be used.

73. Evidence is becoming available (paragraph 6) of structures in the USA, where the requirements for cover are a lot less onerous than in the draft code, that have survived in climates similar to that of parts of the UK for 50 years.

Mr Johnson (Paper 6)

74. In reply to Dr Wyatt, experience has shown that the use of wall ties in grouted cavity work is required even in four-course lifts. The hydraulic head of the wet grout and the disturbance caused when tamping the grout in can often disrupt the masonry, and so the provision of ties is a necessary precaution.

Mr Bradshaw and Mr Drinkwater (Papers 7 and 9)

75. In reply to Mr Astin and Mr Zukas, no special consideration was given to the damp-proof membrane in this particular case since the walls were continuous around the perimeter of the church and hence self-supporting in resisting any horizonal thrust that might arise from arching action. Tension reinforcement at the top of the wall over supports was not considered necessary in this case due to the relatively modest load supported and the high wall-height-to-span ratio (2.8). We would agree that further guidance is needed regarding continuity effects for more

heavily loaded walls.

76. Mr Safier's comments on epoxy-coated reinforcement are of interest. There are many applications for reinforced masonry where the required building life and degree of exposure are such that the cost of stainless steel cannot be justified and yet unprotected steel reinforcement is considered adequate. The cost of epoxy-coated reinforcement lies between the two, along with galvanizing and stainless steel coated, and we await the results of the BRE tests on durability and resistance to mechanical damage with interest.

Dr Curtin, Mr Shaw, Mr Beck and Mr Pope (Paper 8)

77. Dr de Vekey is right in noting that the concrete ring beam bears on the inner leaf and cross-ribs alone and will therefore induce a shear stress at the cross-rib/outer leaf interface. This was catered for by the insertion of shear-ties.

78. In reply to Mr Astin, the stability of diaphragm walls during construction is an important consideration and is usually provided for by suitably braced scaffolding to both faces of the wall. No contractor, so far, has found this to be a problem.

79. In reply to Mr Woolley, the prevention of corrosion of reinforcement in masonry is basically similar to that in concrete (i.e., provide sufficient and adequate cover). We are not happy about galvanizing bed joint reinforcement since this can be too easily damaged on site. Stainless steel should be considered in positions of severe exposure. The protection of vertical reinforcement in hollow infilled blockwork is less of a practical problem on site (as, in general is all reinforced hollow blockwork compared with normal brickwork). The blockwork may be considered as a permanent shutter and, as with in situ concrete placed in timber shutters, the provision of sufficient spacers, ties, care in casting, etc. can ensure a satisfactory result.

PAPERS 10-12

Mr D.P. Moffitt (Mallagh Luce & Partners, Dublin)

80. The lever method of prestressing described in Paper 10 was invented and patented by Mr Mallagh, founder of Joseph Mallagh & Son, Consulting Engineers, Dublin. The firm is now Mallagh Luce & Partners.

81. Five multi-bin silos have been built (Table 2 and Fig. 5) and several other one- or two-bin silos, including one at Greenwich in 1959 which is 25 ft dia. x 90 ft high.

Table 2. Multi-bin silos

	Date	Place	Contents	Capacity	Size
1.	1952	Dublin	Grain	2,000 tons	11 bins, 12 ft dia. x 74 ft high
2.	1954	Enniscorthy	Grain	3,000 tons	12 bins, 12 ft dia. x 93 ft high
3.	1955	Bristol	Grain	3,000 tons	8 bins, 20 ft dia. x 53 ft high
4.	1956	Hull	Grain	3,000 tons	8 bins, 20 ft dia. x 53 ft high
5.	1965	Dublin	Rock phosphate	20,000 tons	10 bins, 30 ft dia. x 80 ft high

Fig.5. 20,000 ton prestressed concrete rock phosphate silo (before steelwork and cladding fixed)

82. The first multi-bin silo has three-sided inter-bin spaces which are filled with mass concrete. All others have four-sided inter-bin spaces which are used for storage except in the rock phosphate silo, where they are empty.

83. In silos 1, 3 and 4, inside and outside scaffolding was used. In silo 2 a platform raised on the blockwork was used instead of inside scaffolding. In silo 5 the scaffolding was inside only with a scaffolding tower between each pair of bins at the jacking point.

84. Straight blocks were used in the first silo and curved blocks subsequently. Straight blocks require a greater thickness of gunite to cover the projecting corners. Curved blocks cost about one third more than straight.

85. The lever stressing method was used in the first four silos. This method was not acceptable to the LCC for the silo at Greenwich and single wire jacks were used for stressing in this silo and the rock phosphate silo built in 1965.

86. Gunite was used on all silos except the last, which was covered with asbestos cement sheeting on a structural steel framework.

87. The rock phosphate silo (excluding plant) cost £7 per ton stored in 1965 which would be about IR£70 today or £60 sterling. The blockwork and stressed wires alone cost £3.75 per square yard in 1965 which would be about IR£37.50 today or £30 sterling.

Mr J.C.M. Forrest (Kenchington Little & Partners, London)
88. This conference is almost entirely devoted to the viewpoint of the professional engineer and the researcher. There are, I believe, no contractors present which is, in my opinion, a great pity for from remarks by Mr Adams and others there is clearly a need to raise standards on site both in craft skills and supervision by contractor's staff.

89. We do appear to have a real problem for there are bags of enthusiasm, inspiration and down-to-earth fun being exercised by my colleagues but no readily recognized organizational ability to deliver the goods on site. Perhaps the expertise does exist but few contractors are laying claims to it.

90. It was my great fortune to start my career in this industry as an apprentice bricklayer which gave me a valuable insight into the large potential of masonry (brick and block) as a material in construction. However, at no time in my training was I introduced to reinforced masonry or prestressed masonry. The City & Guilds syllabus for brickwork did not include any reference to it and I wonder if it does now in the 1980s. Certainly all the test exhibits at national craft competitions show only unreinforced masonry as the height of a brick- or blocklayer's skill. What is needed is surely more education at craft level and supervisory level.

91. With this apparent lack of attention to the training needs for site skills in reinforced masonry and prestressed masonry, may I put a few questions, not solely to the speakers at today's session but to the audience and our two co-sponsoring Institutions.

92. First, what is the viewpoint of the two sponsors of this conference to the site training needs in this technology? Are the sponsors concerned only in the professional development of the technology of reinforced masonry and prestressed masonry?

93. Secondly, what role should the two main trade associations of brick and block, acting in unison, have in providing training of supervision by contractor's staff and training of craft skills?

94. Thirdly, who among speakers and the audience is prepared to devote more thinking and time to the development of

site training in reinforced masonry and prestressed masonry?

95. Fourthly, where are the contractors today who should be reporting on the front-line problems of application of reinforced masonry and prestressed masonry theory and detailing on site?

Rear Admiral A.J. Monk (The Brick Development Association, Windsor)

96. The Association shares Mr Forrest's regret that no contractors were present at the symposium, and it is hoped that there will be more interest among many professional bodies, including contractors, as the techniques for using reinforced and prestressed masonry become better known.

97. BDA agrees entirely with the need for training in site skills both at craft and, most important, at supervisory level. The Association is currently offering training courses specifically to enable contractors' site supervisory staff to obtain a better understanding of all aspects of brick masonry. BDA Training Services Limited is an associated company for training bricklayers and other construction craftsmen. However, craft training is, in the main, a matter for those contracting therein.

98. BDA acts in unison with trade associations for blocks, in areas where interests are common.

Dr R.E. Rowe (Cement and Concrete Association, Slough)

99. The Cement and Concrete Association does provide training courses on most aspects of the uses of cement and concrete and indeed have, for a number of years, been running courses on blockwork. The lectures given on such courses include those on both reinforced and prestressed blockwork. However, although the Association has the facilities to provide the training we do depend on the demand from industry, as a whole, which is in the best position to define its needs. By this I mean that the block industry, architects, consultants and contractors need to identify clearly the types of courses required and, as important, provide participants on them. Thus there is, as Mr Forrest suggests, a need for the industry itself to decide what it most requires in the way of specialized courses and the various levels at which these should be produced, and then to discuss with bodies like the Association the means to meet the need. The Association is only too willing to participate in discussion and to provide relevant courses.

DISCUSSION

Mr R. Milne (Institution of Structural Engineers, London)

100. By the nature of their function, the Institutions of Civil and Structural Engineers must be fundamentally concerned with the development of the technology of reinforced and prestressed masonry. The methods are of relatively recent origin yet the Institutions have published papers and reports, have discussed design applications and, indeed, by this conference itself, have brought together all those who wish to learn and apply the advantages of the techniques.

101. Site training needs are equally important. Few structural designers have the advantage of Mr Forrest of basic training as a skilled craftsman. The construction industry and trades unions, however, through the Construction Industry Training Board, support the CITB Training Centre at Bircham Newton, Norfolk. It is to be hoped that the two trade associations mentioned have, or will, co-ordinate effort/information/know-how to enable the Bircham Newton centre to include practical training for reinforced and prestressed masonry to be included among the craft-skill courses available to personnel sent by their employees (contractors) for training/updating.

Dr D. Lenczner (University of Wales Institute of Science and Technology, Cardiff)

102. My comments concern the loss of prestress in masonry due to creep.

103. The loss of force in the tendons due to the effects of creep in brick masonry may be calculated on the assumption that creep is given by the total elastic deformation of the member under the action of the prestressing force multiplied by an appropriate constant C_c.

104. For members with a ratio of overall length to thickness of 9 or more

$$C_c = 4.46 - 0.33\sqrt{X}$$

and for members with a ratio of overall length to thickness of 1

$$C_c = 1.73 - 0.14\sqrt{X}$$

where X is a coefficient which may be taken to be numerically equal to the strength of the brick unit in N/mm^2. For members with a ratio of overall length to thickness between 1 and 9, linear interpolation is permitted.

105. In the absence of sufficient information for block masonry it is recommended that the value of C_c as calculated above be multiplied by a factor of 2.

106. No information is at present available for brick masonry with calcium silicate bricks.

Dr J.G.M. Wood (Mott, Hay & Anderson, Croydon)

107. Papers 5 and 9 show some limited applications of reinforcement to brick and blockwork silos. In the USA concrete stave silos have been built since the turn of the century and current sales by all producers are typically in the range 5000-10,000 silos per year. Sizes in the USA range up to 9 m dia. and 30 m high, but the largest sizes are not trouble-free. The silos in the 4-8 m dia. range, with heights up to 3 diameters are well proven for a wide range of applications. In agriculture they are used for silage and grains and in industry for coal, cement, aggregates and other powders and granular materials. The same precast units are also used for the construction of water, slurry and latex tanks. There are over 1000 structures of this type in the UK. Their design is covered by BS 5061 for silage silos and ACI-313 for industrial silos.

108. The method of construction has evolved from the wooden barrel. The concrete block stave units, 750mm x 300mm x 75mm, are 'book-ended' to interlock and weigh about 35 kg. A special rig and hoist is used to dry-build the full height silo barrel which is hooped externally with galvanized 14 mm rods. After the joints between staves have been well pointed with mortar, the hoops are post-tensioned. The threaded ends of the hoop sections are lapped as they pass through the 'lug' connectors. The nuts are tightened until the lugs 'turn', as the moment from the eccentricity becomes sufficient to yield the hoop bars under tension and bending. A fully 'turned' lug indicates a tension equivalent to half yield stress in the hoop.

109. For industrial silos the interior may be gunited to increase the wall strength. External vertical tensioned bars are fitted in the UK to resist wind and silage uplift on BS 5061 silos.

110. Because all the reinforcement is external it is inspectable and replaceable. This is particularly important for silage silos, which are vulnerable to acid attack.

111. Over the years people have frequently advocated the use of torque-controlled tensioning in place of the use of the lugs on stave silos. It doesn't work as the friction on threads is too variable, particularly when grit gets on the thread. Experience with torque tightening for HSFG bolts has also been unsatisfactory. I have considerable doubts about the reliability of the torque tensioning method for 'brickwork diaphragm walls' described by Mr Shaw. In his

case great care was obviously taken but not all contractors have finesse. Perhaps tensioning using an initial low torque to bed in followed by a predetermined number of turns of the nut to give the required extension would give a more consistent initial tension.

Mr C. Southcombe (Plymouth Polytechnic)

112. Mr Haseltine says that it has long been accepted that reinforced masonry design can be based on reinforced concrete principles. Work carried out at Plymouth has indicated that this statement is not entirely valid. Tests carried out on reinforced brick beams and brick panels with bed joint reinforcement indicate that there is a greater reserve of strength in these elements than is shown in a reinforced concrete analogy. In particular, tests on beams (36 in number) at Plymouth, using different spans and bricks and having the format shown in Fig. 6, lead to consideration that

(a) there is a useful working tensile force in brickwork
(b) shear presents no problems if a ceramic bond is used

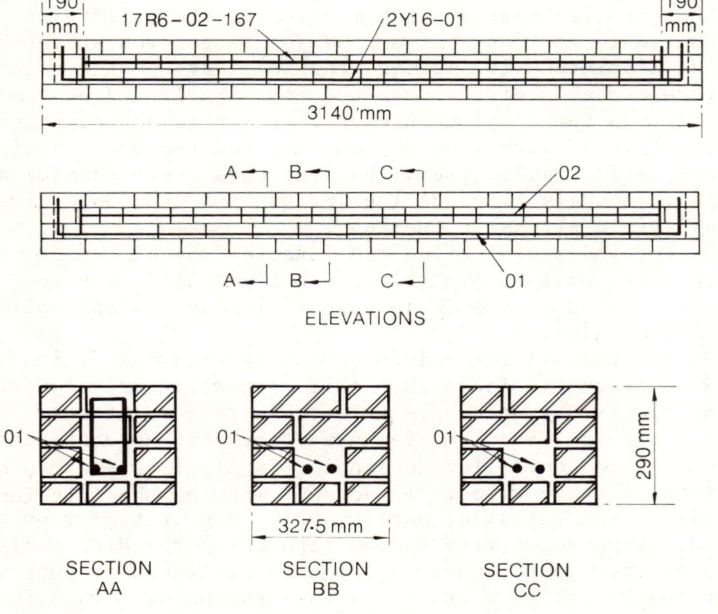

Fig.6. Test beams

(c) if a beam is designed to satisfy serviceability conditions (i.e., deflexion) then cracks are not particularly significant

(d) the grout infill used, which was of relatively low strength, did not affect the strength of the beams.

113. A comparison of cost of a reinforced concrete beam and similar reinforced brickwork beam showed a saving if the latter was used.

114. Compressive failures have been achieved in tests at Plymouth, and beams tested have satisfied design criteria of balanced section, over-reinforced section and under-reinforced section; that is, tests have been designed to initiate plastic failure, brittle failure and simultaneous plastic and brittle failure.

Dr J.A. Purkiss (Department of Civil Engineering, University of Aston)

115. I am afraid that in order to deal adequately with structural analysis and design in steelwork and concrete in an undergraduate course, masonry is treated as a peripheral subject. This is a regrettable state of affairs and may have contributed to its lack of use.

Mr Mallagh (Paper 10)

116. Among the papers submitted to this symposium and the contributions to this part of the discussion, only two projects dealt in any detail with the practical application of prestress to masonry construction: stave silos in the USA described in the contribution by Dr Wood and the prestressed blockwork silos covered in my own paper. While the number of stave silos mentioned by Dr Wood, 5000-10,000 per year, is most impressive, their size appears to be considerably smaller than those for which Joseph Mallagh & Son (now Mallagh Luce & Partners) have been responsible; also the former seem to be mostly single-bin units while the latter are generally multi-bin with the inter-bin space used for storage.

117. As mentioned by Mr Moffitt (paragraph 85), the LCC (now GLC) would not accept the lever method of prestressing as they considered it was not sufficiently positive. This was responsible for the change to jacking, with the advantages as described in my paper.

118. Dr Wood's contribution does not make clear how the load in the 14 mm circumferential rods was found to be the required amount. He refers to lugs turning. Presumably the lugs were proportioned so that they would bend when the

requisite load was reached in the 14 mm rods. It is doubtful if the GLC would consider this a sufficiently positive method of determining prestressing load.

119. When the lever method of prestressing was being developed, another engineer advocating circumferential bars for prestressing silos suggested that they should be levered off the silo wall until the millscale flaked off them - again a far from positive method of determining load in a tendon.

120. A point of significance between the stave and the prestressed block silos is that the joints in the former were merely pointed and in the latter a high quality mortar was used for jointing. Creep resulting in loss of stress would be more likely to occur with pointed joints. With the blockwork silos particular attention was paid to quality of block and joint material to keep to a minimum loss of stress due to creep. Strain gauge measurements on the circumferential wires of silo 5 (paragraphs 80-87), carried out 12-20 months after stressing, indicated that loss of stress due to creep was negligible.

Mr Shaw (Paper 11)

121. With any structural system there is a need for the engineer to be aware of the critical aspects of construction as well as any long term changes to the system behaviour.

122. In the case of torqued up bars in diaphragm walls the engineer's site inspections should check that there are no obstructions which could affect the force being applied and that the threads are clean and lightly oiled. This is particularly important at the time of applying the post-tension force. Such supervision checks on the ungrouted post-tension rod inside the large void of the diaphragm is relatively simple.

123. I agree with Dr Wood that the relationship between the amount of torque applied and the tension developed in the rod is not accurate but I do not expect it to be. Practical engineering is not accurate in any material or system. Construction and material variations can and will occur and the engineer must understand this and make due allowance for it.

124. Some of these variations will mean that the full tension calculated will not necessarily be produced in the bar and I agree that long term losses will also deduct some tension. But this system does not only produce errors which reduce the tension; for example, tests have shown that even if the nut on the capping plate is only finger-tight the rod will pick up tensile stress as the lateral load is applied. This increase is caused by the eccentricity of the post-

tension rod and increases the lateral resistance of the wall. It is also known that clay bricks expand with time and this also increases the tension in the rods.

125. Engineering is applying a practical balance of errors when making assumptions in order to achieve a successful result. Post-tensioned masonry is no more difficult or more critical than any other solution to engineering problems and in fact if understood and properly detailed it can be made very simple and reliable and we have many projects that prove it.

Mr Haseltine (Paper 12)

126. Dr Lenczner's work will be useful in improving the draft of Part 2 of BS 5628 but I would suggest, after looking at his information, that the two formulae could be simplified to

$$C_c = 4.5 - 0.3\sqrt{X}$$

and

$$C_c = 1.7 - 0.2\sqrt{X}$$

127. I am grateful to Mr Southcombe for the information on his tests. If he has found that the use of reinforced concrete design methods is conservative, the drafting committee may be able to improve the formulae in the draft Part 2 of BS 5628. However, it has always been the practice to ignore any contribution from the tensile strength of the masonry, as is also done in reinforced concrete design. I hope that the detailed results, or at least a discussion on them, will be made available to the BSI Committee CSB 33/2 for use in drafting Part 2.

REFERENCE

1. TOVEY A.K and ROBERTS J.J. Interim design guide for reinforced concrete blockwork subject to lateral loading only. Cement and Concrete Association, London, 1980, interim technical note 6.

Closing address

C. J. EVANS, President-Elect, Institution of Structural Engineers

This conference has been extremely valuable in providing a forum for the exchange of information on practical experience of reinforced and prestressed masonry, a subject on which information is somewhat limited in the UK. This is not the case in other countries, where masonry has been used as a structural material for many years; it is not therefore a new material, although it may appear to be so in the UK, but rather a resurgence of a material, with obvious advantages.

One of the main advantages of masonry is its long life, which has not yet been proven with, say, reinforced concrete, simply because reinforced concrete has not yet been in existence long enough. However, as many contributors to the conference have asserted, further research is necessary, particularly in the fields of corrosion and creep, if the objective of the long life of masonry is to be achieved in reinforced and prestressed brickwork. I believe that there is a need for greater co-operation between the research worker and the practising engineer, and in addition to pure research being done in the fields of corrosion and creep, it is important to monitor actual performance in actual structures. The problem is, who pays? Clients are not likely to, and moreover might become worried if permission were requested to monitor the performance of their structures. However, this is a problem that must be faced, as I believe that monitoring of structural performance is as important as pure research.

Another problem that has been mentioned is the reluctance of contractors to price reinforced and prestressed masonry realistically, because of the involvement of two separate trades. This is surely a problem that can be overcome, and I am pleased to learn that discussions are taking place between the relevant Associations in an effort to overcome this.

The reluctance of engineers to design in a material with which they are not familiar, has also been mentioned as a reason that reinforced and prestressed brickwork is not more widely used. This is surely a problem for our universities to deal with to ensure that engineering graduates have at least some knowledge of the principles of the design of reinforced and prestressed brickwork. Perhaps also the publication of the code of practice on structural masonry will add some respectability to this material in the minds of designers, and will give them more confidence to design using structural brickwork.

One of the problems continually facing the engineering profession is the dissemination of information, both of research developments and, particularly, of practical experience. This conference has been an excellent example of the way in which practical experience can be passed on to others, and I have no doubt that a further conference at some time in the future, when further experience has been gained and needs to be passed on, will be equally beneficial.

Bibliography

Prepared by the Cement and Concrete Association and Brick Development Association

REINFORCED AND PRESTRESSED MASONRY

This bibliography covers the principle publications. Further information is however available on this and unreinforced masonry for structural and general use.

BOOKS AND DESIGN GUIDES

CURTIN, W.G., SHAW, G., BECK, J.K., BRAY, W.A. Structural Masonry Designers' Manual. St Albans, Granada, 1982. 512 pp (to be published May 1982)

This book covers the structural elements and forms of brick and blockwork. For each element or form the book gives designs, applications, structural techniques as well as problems and solutions. Wherever possible points are illustrated with a diagram or table and step-by-step design examples of typical elements and buildings. Design features include crosswall and cellular construction, masonry spine walls, arches, the applications of reinforcing and post-tensioning etc. Special features of this book are the authors' own innovative structural developments. These are given throughout.

ROBERTS, J.J., TOVEY, A.K., CRANSTON, W.B., BEEBY, A.W. Concrete Masonry Designer's Handbook. London, Eyre & Spottiswoode. (to be published mid. 1982.)

This handbook deals with the general topic of concrete masonry from materials and physical properties through to structural design and construction. It has been written as a text book in that it covers background information to BS 5628 together with research and design data. There are comprehensive chapters dealing with the design of unreinforced and reinforced masonry in limit state terms and chapters dealing with non-structural aspects such as fire, thermal, sound, movements, specifications, etc. Design examples are given throughout together with tables, graphs and other design aids.

HENDRY, A.W. Structural Brickwork. London. Macmillan, 1981. 200 pp

A review of the theoretical basis of structural brickwork related to practical design. Each chapter has an extensive list of references making it possible to trace source material. "Reinforced and Prestressed Brickwork", and "Brick Masonry Walls in Composite Action" are but two of the eight chapters.

BIBLIOGRAPHY

TOVEY, A.K., ROBERTS, J.J. Interim Design Guide for Reinforced Concrete Blockwork subject to lateral loading only. London. Cement & Concrete Association. 1980. 44 pp. Publication ITN.6.

The main purpose of this interim technical note is to provide a guide to the design of reinforced concrete blockwork subject to lateral loading only, pending the publication of Part 2 of BS 5628 "Code of Practice for the structural use of masonry", which will cover the use of reinforced and prestressed masonry. The note reviews the design methods currently available and selects a limit-state method on the basis of research carried out at the Cement & Concrete Association. Design requirements are summarized, and aid is given in the form of design charts. Detailed calculations of a typical example are given in an appendix, together with loading data for stored materials.

BRICKWORK RETAINING WALLS. Brick Development Association, 1979. Publication D.G.2.

Design Guide for mass and reinforced brickwork retaining walls in both permissible stress and limit state terms.

REINFORCED BRICKWORK RETAINING WALLS. Structural Clay Products, 1979. Publication SCP 15.

A detailed design guide with design tables for walls and bases including a sample calculation. Adopts the limit state approach. A useful design tool.

REINFORCED BRICKWORK POCKET TYPE RETAINING WALL. Structural Clay Products, 1977. 13 pp. Publication SCP 13.

A publication describing the design and construction of a reinforced brickwork retaining wall to meet a given set of conditions.

DESIGN GUIDE FOR REINFORCED AND PRESTRESSED CLAY BRICKWORK. British Ceramic Research Association, 1977. Publication SP 91.

A well illustrated and comprehensive document giving the limit state design procedure for various conditions of reinforced and prestressed brickwork, information on materials and their properties and guidance on the supervision and control on site.

BIRD, A.B. and FOSTER, D. SCP 12 Windmill Tower in Reinforced Blockwork. Structural Clay Products, 1976.

Describes the design of a tall slender reinforced brick windmill tower.

MAURENBRECHER, A.H.P., BIRD, A.B., SUTHERLAND, R.J.M. and FOSTER, D. SCP 10 and 11 Vertical Cantilever Walls In Reinforced Brickwork Volumes 1 & 2. Structural Clay Products, 1976.

Volume 1 (SCP 10) shows that useful structural performance from reinforced brickwork may be obtained using the methods of CP111 : 1970, but tests described in detail in Volume 2 (SCP 11) showed that this Code is excessively conservative. Lateral loading tests were carried out on one prestressed and five reinforced forms of brick retaining wall to determine stiffness, strengths and modes of failure of the walls.

DESIGN OF A PRESTRESSED BRICKWORK WATER TANK. Structural Clay Products, 1975. Publication SCP 9

The design and detailing of a 543 m^3 circular storage tank.

SUTER, G.T., and HENDRY, A.W., Limit State Shear Design of Reinforced Brickwork Beams, Proc. British Ceramic Society 24(1975) 191-6.

Examines the effect of ratio of shear span to effective depth, the amount of tensile reinforcement and brickwork compressive strength on the ultimate shear resistance of reinforced brickwork beams and proposes limit state shear stress values.

THORLEY, W. Design of Loadbearing Brickwork in SI and Imperial Units. London, Heinemann, 1970. 187 pp.

The book deals with practical examples designed to CP 111 : 1964 in both Imperial and Metric Units. The chapter headings include reinforced columns, reinforced beams and reinforced retaining walls.

CHRISTIE, B.E., ISAACS, H.P. Australian Concrete Masonry Design and Construction. Concrete Masonry Association of Australia, 1977.

This publication covers general design and use of both unreinforced and reinforced concrete masonry as used in Australia. The publication has several chapters dealing with such topics as types of masonry units, specifications, properties, mortar, reinforced and unreinforced masonry, applied finishes and construction details and practice.

AMRHEIN, J.E. Reinforced Masonry Engineering Handbook. Los Angeles Masonry Institute of America, 1972. 320 pp.

This book contains comprehensive information on the design and use of reinforced masonry in America. A number of coefficients, tables, charts and design data is presented in an aid to eliminate repetitious and routine calculations. The allowable stresses and design requirements relate to the Uniform Building Code, 1970 Edition, as published by the International Conference of Building Officials. In addition to this basis of design, sound engineering practice has been included

RESEARCH/OTHER PUBLICATIONS

REINFORCED AND PRESTRESSED MASONRY SYMPOSIUM PAPERS. London. Institute of Structural Engineers, 1981. 72 pp.

These papers relate to the Symposium held on 8 July 1981 on the draft Code BS 5628 : Part 2. They cover: Historical background and introduction to BS 5628 : Part 2; The drafting of BS 5628 : Part 2; The approach to bending; The shear behaviour of reinforced concrete blockwork beams; The shear strength of reinforced brickwork; Reinforced masonry subject to combined loading; Practical applications of reinforced and prestressed masonry; A view on the shape of the stress block and material safety factors; Durability of reinforced masonry; Aspects of durability of clay brickwork; Interim results from an investigation of the durability of reinforcing steel in reinforced concrete blockwork.

BARNES, M.M. Farm Note No.11 Concrete Block Walls. London. Cement & Concrete Association, 1981. 8 pp. Publication 47.611.

Explains the construction of a concrete block wall either as part of a building or as a free-standing structure. Advice is included on the construction of reinforced retaining walls.

SUTHERLAND, R.J.M. Brick and Block Masonry in Engineering. Proceeding of Institution of Civil Engineers, Part 1. February 1981.

A review paper surveying the properties of the basic materials and forecasting developments in the use of brick and block masonry in civil engineering, concludes that reinforcing and pre-stressing techniques are likely to be increasingly used.

BIBLIOGRAPHY

ROBERTS, J.J. The development of an electrical resistance technique for assessing the durability of reinforcing steel in reinforced concrete blockwork. London. Cement & Concrete Association, 1980. 19 pp. Publication 42.532.

Tests have shown that an electrical-resistance tecnique is suitable for determining the amount of corrosion that might take place on the steel in reinforced masonry. Allowance must be made for the influence of temperature upon the gauge resistance but the moisture content of the surrounding concrete has little effect.

ROBERTS, J.J. Further work on the behaviour of reinforced concrete blockwork subject to lateral loading. London. Cement & Concrete Association, 1980. 43 pp. Publication 42.531.

The results of tests on the behaviour of vertically reinforced concrete blockwork subject to lateral loading were reported in Technical Report 42.506. These results have since been supplemented by tests on ten horizontally reinforced wall elements and eight vertically reinforced sections. The later results support the conclusion of the earlier report that the present design recommendations for reinfoced concrete blockwork are conservative and uneconomic especially where a high percentage of reinforcement is used.

RATHBONE, A.J. The Behaviour of Reinforced Concrete Blockwork Beams. London. Cement & Concrete Association, 1980. 23 pp. Publication 42.540.

Twelve single-course reinforced concrete blockwork beams with various sizes and arrangements of reinforcement have been tested in flexure. The blocks were dense aggregate, hollow, facing blocks from one manufacturer of one strength, and were 390 x 190 x 190 mm thick. The experimental performance of the beams under two-point flexural loading was compared with predictions calculated by using current design procedures. The tests indicated that ultimate limit state analysis in flexure and the serviceability limit state analysis of deflection are accurate. However, because deep beam action may occur over openings in walls, the performance of such beams with respect to shear and cracking needs further study.

SINHA, B.P. Long Term Tests on Reinforced Clay Brick Masonry Grouted Cavity Vertical Cantiliver Walls. Structural Clay Products Limited, Hertford, 1979. Publication SCP 14.

Describes long term creep behaviour of 2.5 m grouted cavity vertical cantilever walls subject to lateral pressure for one year, at the end of which tests to failure were undertaken and ultimate loads determined

REINFORCED BRICKWORK BOX BEAMS. Structural Clay Products, 1979. 33 pp. Publication SCP3.

A publication which describes experimentation of the use of reinforced brickwork with thin reinforced concrete slabs to form storey height beams and cantilevers.

SINHA, B.P. Reinforced Brickwork: Grouted Cavity Shear Tests. Structural Clay Products Limited, Hertford 1978. Publication SCP 16.

The effect of shear span/effective depth ratio on the ultimate shear strength of reinforced brick grouted beams having a constant percentage of steel was investigated and it was found that ultimate stress increased with decrease in shear span/depth ratio as predicted.

POWELL, B. A Review of the Literature on Reinforced Brickwork. British Ceramic Research Association Technical Note 283, Stoke-on-Trent, 1978

Outlines the history of reinforced brickwork and gives design examples. Comprehensively reviews investigations in many countries determine structural performance of reinforced brickwork.

CRANSTON, W.B., ROBERTS, J.J. The Structural Behaviour of Concrete Masonry - Reinforced and Unreinforced. The Structural Engineer, Volume 54, No.11, 1976. pp 423-436.

In this paper the behaviour of reinforced and unreinforced masonry is considered. Tests on eccentrically-loaded unreinforced walls and couplet specimens are reported and a simple theoretical approach for solid block masonry is derived. A method of predicting wall strength is presented. Additional tests are described on reinforced concrete sections subjected to lateral loading only and the simple ultimate load theory used for reinforced concrete is shown to give a good indication of the ultimate strength of the sections. The effect of employing different values of the partial factor of safety for strength of the masonry is considered. It is indicated that present design procedures using permissible stresses result in uneconomic design.

CADJERT, A. and LOSBERG, A. Lateral Strength of Reinforced Brick Walls. Design for Wind Loads. Proceedings of the Fourth International Brick Masonry Conference (Brugge) 1976, Paper 4.c.4.

Describes lateral loading tests on half brick thick walls with and without reinforcement, and offers some preliminary design rules for reinforced brick panels subjected to wind loading.

SUTER, G.T. and KELLER, H. Shear Strength of Grouted Reinforced Masonry Beams .Proceedings of the Fourth International Brick Masonry Conference (Brugge) 1976, Paper 4.c.2.

Describes experiments to determine the effect of the ratio of shear span/effective depth on the shear strength of grouted reinforced masonry beams and concludes that ultimate shear stress decreases markedly with decreasing ratios of shear span/effective depth.

ROBERTS, J.J. The Behaviour of Vertically Reinforced Concrete Blockwork subject to Lateral Loading. London. Cement & Concrete Association, 1975. 13 pp. Publication 42.506.

Twenty walls constructed from hollow blocks with vertical reinforcement to form panels 1½ blocks wide by 15 courses high (approximately 0.6 x 3 m) were tested as simply supported beams. The results indicate that a limit-state analysis similar to that used in CP 110 satisfactorily predicts the ultimate moment carried by each wall. Furthermore, if the requirements of CP 110 are applied to obtain suitable working moments, it is apparent that, by comparison, design in terms of CP 111 appears to be conservative and uneconomical, especially where a high percentage of reinforcement is used. From an analysis of the measured deflections for the reinforced blockwork walls has been derived.

CURTIN, W.C., ADAMS, S. and SLOAN, M. The Use of Post-Tensioned Brickwork in the S.C.D. SYSTEM. Proceedings British Ceramic Society 24 (1975) 233-45

Describes a system of building for school buildings using post-tensioned brick spandrel panels below structural timber glazing.

THONGCHAROEN, V. and DAVIES, S.R. The Composite Action of Simply Supported Reinforced Brickwork Walls and Reinforced Concrete Beams. Proceedings of Third International Brick Masonry Conference. Edited by L Foertig and K Gobel, Bonn, Bundesverband der Deutschen Ziegelindustrie, 1975. pp 291-7.

Gives the results of tests on mould wall-beams using small scale bricks carried out to study the effect of composite action of wall and reinforced concrete beams where the beam width was equal to or greater than wall thickness and to determine the ultimate load capacities of such constructions.

SUTER, G.T. and HENDRY, A.W. Shear Strength of Reinforced Brickwork Beams. Structural Engineer 53 (1975) 249-53.

FOSTER, D. and THOMAS, A. Some Interim Comments on the Corrosion of Reinforcement in Brickwork. Proceedings British Ceramic Society (24) September 1975.

Reports a small scale investigation by exposure panels with interim conclusions of the effect of depth of cover and different methods of protecting the steel. Also presents observations on demolished r.b. structures.

ABEL, C.A. Prefabricated Reinforced Brick Masonry Arch Bridge. Proceedings of the Third International Brick Masonry Conference (Essen) 1973 ed. L Foertig and K Gobel (Bundesverband der Deutschen Ziegelindustrie, Bonn, 1975) pp. 544-8

Describes the design, construction and erection of a prefabricated parabolic arch bridge in reinforced brick masonry for light vehicular traffic.

ARMSTRONG, A.C. and HENDRY, A.W. The Compressive Strength of Brickwork and Reinforced Bed Joints. British Ceramic RA Technical Note 209, 1973.

Investigation of the strength of brickwork prisms reinforced in the bed joints, which showed that the increase in compressive strength obtained was related to the surface area and number of reinforcing bars used.

MEHTA, K.C. and FINCHER, K. Structural Behaviour of Pretensioned Prestressed Masonry Beams. SIBMAC Proceedings. Edited by H.W.H. West and K H Speed. Stoke-on-Trent. British Ceramic RA, 1971. pp 215-9.

Describes loading tests on prestressed brick beams with various coursing patterns made from a U shaped masonry shell, prestressing the strands and filling the cavity with grout.

GROGAN, J.C. Miscellaneous Reinforced Brick Masonry Structures. Proceedings of the Second International Brick Masonry Conference (Stoke-on-Trent) 1971. Edited by H W H West and K H Speed (British Ceramic Research Association, Stoke on Trent, 1971) pp. 327-30.

Describes various reinforced brick masonry structures in U.S.A. with some discussion on costs and building code requirements.

DINNIE, A. and BEARD, R. Reinforced Brickwork Silos for Grain Storage. Proceedings British Ceramic Society 17 (February 1970) 121-135.

Details of the design and construction of two forms of reinforced brickwork for the bulk storage of grain. In one form, radial bricks are used for circular silos with walls a half brick thick, in the other, standard bricks are used in quetta board for rectangular silos with walls one and a half bricks thick.

PLOWMAN, J.M., SUTHERLAND, R.J.M. and COUZENS, M.L. The Testing of Reinforced Brickwork and Concrete Slabs Forming Box Beams. Structural Engineer 45 (11) November, 1967

The paper describes a series of full-scale tests on composite cantilever box beams having reinforced concrete slabs as flanges and reinforced brick walls as webs.

CLAY PRODUCTS TECHNICAL BUREAU. Post-Tensioned Brickwork and its Use in the Construction of a Factory at Darlington. C.P.T.B. Technical Note 1 (9), May, 1966

Describes the use of post-tensioned brickwork in a factory building having 24 ft high cavity brick walls.

RECENT STRUCTURAL AND GENERAL PUBLICATIONS ON MASONRY

STRUCTURAL

CURTIN, W.G. SHAW, G. BECK, J.K. BRAY, and BRAY, W.A. Design of Brick Diaphragm Walls. March 1982. Brick Development Association. 41 pp.

Authoritative design guidance on this efficient and economical form of brick construction. Discusses architectural opportunities and engineering design principles. Summarises design procedure and includes worked examples.

CURTIN, W.G. SHAW, G. BECK, J.K. and BRAY, W.A. Crosswall Construction for Commercial Buildings. Brick Development Association. (To be published 1982)

Design principles and guidance for the crosswall construction technique.

CURTIN, W.G. SHAW, G. BECK, J.K. and DRAY, W.A. Reinforced Brickwork. Brick Development Association. (To be published 1982)

A comprehensive treatment of reinforced brickwork with design examples.

HAZELTINE, B.A. and MOORE, J.F.A. Handbook to BS 5628 : Structural Use of Masonry. Part 1 : Unreinforced Masonry. The Brick Development Association, 1981. 118 pp.

This handbook includes detailed discussion of every clause in the Code, supported by references from the now very extensive literature on masonry construction and by a large number of illustrative examples. The latter are presented in full numerical detail, and cover practically all likely design problems. The handbook is authoritative on the basis of the authors' direct knowledge of the subject and of the discussions in the drafting committee. It is also practical because of their extensive experience in masonry design, construction and research. Users of BS 5628 will be much indebted to the authors of the handbook for their work in clarifying the basis of the code and its application to practical design.

FISHER, B.H. Guide to the structural design of precast concrete blockwork in accordance with BS 5628 : Part 1 : 1978. Aggregate Concrete Block Association/Autoclaved Aerated Concrete Products Association, 1981. 23 pp. Publication 73.312.

The Design Guide is intended to help suitably qualified designers use the British Standard BS 5628 : Part 1, taking full advantage of the range of thicknesses and strengths of precast concrete blocks which are available. It also gives guidance on the measures to be adopted to avoid collapse due to accidental damage, and makes reference to relevant publications relating to other aspects of the design of blockwork.

BIBLIOGRAPHY

HENDRY, A.W. SINHA, B.P. and DAVIS, S.R. An Introduction to Loadbearing Brickwork Design. Ellis Horwood Limited, 1981.

A student and mid-career training textbook for engineers, incorporating the latest information available regarding limit state design based upon BS 5628. Includes design calculations for a seven-storey dormitory building.

BUILDING RESEARCH ESTABLISHMENT. Strength of brickwork and blockwork walls : design for vertical load. Digest 246. Garston, Building Research Station, 1981. 8 pp.

This digest gives some background to the main recommendations in respect of vertical loading and replaces Digest 61.

CURTIN, W. etal. Design of brick walls in tall single storey buildings. Brick Development Association. 1980. 32 pp. Publication DG8.

Discusses the development of fin wall construction, architectural opportunities, engineering design principles, and design method. Provides detailed design examples, and recommended design procedure.

LENCZNER, R.D.L. Brickwork - Guide to Creep. Structural Clay Products, 1980. pp 26. Publication SCP17

This note gives a number of guidelines for the calculation of elastic and creep movements in brickwork walls and piers subjected to sustained axial loads. They are based on 12 years research by the author and his research team and on the most up to date information.

CURTIN, W. Brick diaphragm walls in tall single storey buildings. Brick Development Association, 1977. Publication DG3. Revised edition 1979.

Authoritative design guidance on this efficient and economical form of brick construction. Discusses architectural opportunities and engineering design principles.

HASELTINE, B.A. & TUTT, N.J. External walls design for wind loads. Brick Development Association, 1978. 28 pp. Revised edition 1979.

Provides guidance on the design of solid, cavity and free-standing walls subjected to lateral forces. Contains example calculations using both the premissible stress and limit state design methods.

READ, J.B., CLEMENTS, S.W. The strength of concrete blockwalls. Phase III : Effects of workmanship, mortar strength and bond pattern. London. Cement & Concrete Association, 1977. 10 pp. Publication 42.518.

This report covers the third phase of the experimental investigation into the loadbearing characteristics of concrete masonry. The effects of workmanship, mortar strength and type, and bond pattern of a particular type of 200mm thick block are studied. The report concludes that, of the variables investigated for walls under uniaxial load, bond of the 200 mm hollow block is significant and the workmanship and mortar strength have a lesser effect.

GENERAL

TOVEY, A.K. Concrete Masonry for the Designer. London. Cement & Concrete Association, 1981. 20 pp. Publication 48.049.

Intended for designers and specifiers, the booklet describes the major features of concrete blocks and bricks, their physical properties, design aspects in relation to the new British Standard for concrete masonry units (BS 6073) and other features essential for successful design

TOVEY, A.K. Concrete Masonry for the Contractor. London. Cement & Concrete Association, 1981. 20 pp. Publication 48.050.

Intended to give guidance to the contractor on the principles of good practice and workmanship relevant to the construction of concrete masonry. Outlines the important features of concrete blocks and bricks in relation to the new British Standard for these units (BS 6073) and deals with practical details such as the provision of movement joints, blocklaying techniques, and reinforced masonry construction.

A GUIDE TO THE NEW STANDARD FOR CONCRETE BLOCKS. Leicester. Aggregate Concrete Block Association, 1981. 2 pp. Publication 73.310.

This leaflet gives information on aggregate concrete blocks to the new British Standard for precast concrete masonry units, BS 6073, which replaced BS 2028, 1364 and BS 1180 during 1981.

SP 56 : 1980. Model Specification for Clay and Calcium Silicate Structural Brickwork. British Ceramic Research Association. Distributed by Brick Development Association.

A most important document, providing guidance for the specification of brickwork.

TOVEY, A.K. Model Specification for Concrete Blockwork. London. Cement & Concrete Association, 1976 (revised edition due mid.1982.) 12 pp. Publication 48.043.

Intended to help designers prepare suitable specifications for traditional concrete blockwork. Gives recommended clauses, accompanied by notes for guidance, for the masonry materials and workmanship, and for related work such as flashing, damp-proofing, plastering and rendering.

BRITISH STANDARDS AND CODES

BS 5628 : Part 1 : 1978	Structural Use of Masonry - Unreinforced
BS 5628 : Part 2	Structural Use of Masonry - Reinforced and Prestressed (currently with BSI committee considering public comments)
CP 111 : 1972	Structural Recommendations for loadbearing walls
CP 121 : 1973	Code of Practice for Walling (under revision, eventually to be published as BS 5628 : Part 3)
BS 187 : 1978	Calcium Silicate - sand lime and flint lime - bricks
BS 3921 : 1974	Bricks and Blocks of Fired Brickwork, Clay or Shale (under revision)
BS 6073 : 1981	Precast Concrete Masonry Units.